# Regenerating forests and livelihoods in Nepal

# - A new lease on life

*Unfolding the experience of 20 years of poverty alleviation through leasehold forestry in the Himalayas*

**Authors:**

Bala Ram Adhikari – Department of Forests, Government of Nepal

Nav Raj Baral – Forestry specialist

Jim Hancock – Natural resources Officer, Food and Agriculture Organization (FAO-UN), Rome

Govinda Kafley – Team Leader, TA-LFLP

Pashupathi Koirala – Department of Forests, Government of Nepal

Jaap Reijmerinck – Agriculture and rural Economist

Brett Shapiro, lead author, technical writer

**Under the coordination of:**

Benoit Thierry, Country programme manager, International Fund for Agriculture Development (IFAD), Rome

**Editors:**

Kenichi Shono, Forestry Officer, FAO Regional Asia Pacific Office (FAO RAP)

CABI, UK

# REGENERATING FORESTS AND LIVELIHOODS IN NEPAL: A NEW LEASE ON LIFE

*UNFOLDING THE EXPERIENCE OF 20 YEARS OF POVERTY ALLEVIATION THROUGH LEASEHOLD FORESTRY IN THE HIMALAYAS*

Published by
The Food and Agriculture Organization of the United Nations (FAO)
and
International Fund for Agricultural Development (IFAD)
by arrangement with
CABI

Food and Agriculture Organization of the United Nations

IFAD
Investing in rural people

www.cabi.org

CABI
Nosworthy Way
Wallingford
Oxfordshire OX10 8DE
UK

Tel: +44 (0)1491 832111
Fax: +44 (0)1491 833508
E-mail: cabi@cabi.org

CABI
745 Atlantic Avenue
8th Floor
Boston, MA 02111
USA

Tel: +1 617 682 9015
E-mail: cabi-nao@cabi.org

Website: www.cabi.org



ISBN: 978 1 78064 631 2 (CABI hardback edition)
ISBN: 978 1 78064 632 9 (CABI paperback edition)

ISBN: 978-92-5-108898-2 (FAO paperback edition)
ISBN: 978-92-5-108899-9 (FAO hardback edition)

Photo Credits: ©LFLP - FAO - IFAD
Photo Archives, LFLP staff, 1996-2014
Kaushal Shreshtra 2012-2014
Benoît Thierry, 2011-2014

A catalogue record for this book is available from the British Library, London, UK.

**Library of Congress Cataloging-in-Publication Data**

Names: Thierry, Benoit.
Title: Regenerating forests and livelihoods in Nepal : a new lease on life.
Description: Boston, MA : CABI, 2015. | Includes bibliographical references and index.
Identifiers: LCCN 2015029263| ISBN 9781780646312 (hbk : alk. paper) | ISBN 9781780646329 (pbk : alk. paper)
Subjects: LCSH: Community forestry--Nepal. | Forest regeneration--Nepal. | Rural development--Nepal.
Classification: LCC SD657..N4 T45 2015 | DDC 634.9/56095496--dc23 LC record available at http://lccn.loc.gov/2015029263

Commissioning editor: Caroline Makepeace
Editorial assistant: Emma McCann
Production editor: Tracy Head

Printed and bound in the UK from copy supplied by the authors by CPI Group (UK) Ltd, Croydon, CR0 4YY.

## Table of Contents

# Acronyms

| | |
|---|---|
| BISEP-ST | Biodiversity Sector Program for Siwaliks and Tarai |
| CBFM | Community Based Forest Management |
| CFUG | Community Forest User Group |
| DFID | Department for International Development |
| DFO | District Forest Office/Officer |
| DoF | Department of Forest |
| Ex-ACT | EX-Ante Carbon balance Tool |
| FAO | Food and Agriculture Organisation of the United Nations |
| GDP | Gross Domestic Product |
| Ha | Hectare |
| HLFFDP | Hill Leasehold Forestry and Forage Development Project |
| ICIMOD | International Centre for Integrated Mountain Development |
| IFAD | International Fund for Agricultural Development |
| IPS | Inter Press Service |
| LFLP | Leasehold Forestry and Livestock Programme |
| LFP | Livelihoods and Forestry Programme |
| LFUG | Leasehold Forestry Users Group |
| MAPs | Medicinal and Aromatic Plants |
| MFSC | Ministry of Forest and Soil Conservation |
| MSFP | Multi Stakeholder Forestry Program |
| NGO | Non-governmental Organization |
| NPC | National Planning Commission |
| NTFP | Non-Timber Forest Product |
| REDD+ | Reduce Emissions from Deforestation and Forest Degradation |
| SNV | Netherlands Development Organisation |
| TA | Technical Assistance |
| UN | United Nations |
| UNDP | United Nations Development Programme |
| WUPAP | Western Upland Poverty Alleviation Project |

Currency conversion:
1 USD (US Dollar) = 100 NPRs (Nepalese Rupees)

# List of tables, charts and pictures

## Tables



## Figures

# Preface

"Regenerating forests and livelihoods in Nepal - A new lease on life", recounts a success story about a ground-breaking and innovative rural development operation in the Himalayan foothills that made a real difference: during a lifespan of 20 years, it transformed increasingly degraded land, where farmers carved out a meager existence, into a fertile and prosperous zone where they could thrive. The ingredients to the programme's success, from both technical and human perspectives, are described in this book and demonstrate people's extraordinary commitment in this country prone to natural disasters, particularly earthquakes. This positive experience deserves to be shared in this time of major global commitment to environmental and climate change management

Spearheaded under the Roof of the World, the Leasehold Forestry and Livestock Programme (LFLP) teams worked closely with the local inhabitants, civil society and the government. The tasks ranged from bringing in cultivation techniques that protected the environment, to introducing new local crops and goats, to promoting an entrepreneurial spirit and marketing expertise. Every intervention was carefully conducted to mitigate climate change and to help most vulnerable farmers become more resilient to its effects as well as improving their family income.

Farmers who took part in the Leasehold Forestry programme speak candidly about how the initiative helped them to increase their production of tree crops and livestock without degrading the precious forest upon which their livelihoods depend. Time and time again, their testimonies bear witness to the importance of "inclusiveness" – ensuring that everyone in their community, especially women and young people, take part in every aspect of the learning and the doing.

The forest was allocated to communities with 40-year renewable leases and then restored by the Leasehold Forestry programme staff and villagers to create viable and sustainable incomes and reduce poverty. The result? The dramatic "greening" of once barren lands; abundant harvests and new products; a higher standard of living – in short, a better life and one that holds great promise for the generations to come.

Although aid development initiatives produce thousands of documents, the valuable information they contain remains buried in bureaucracies and in the "grey literature", never to surface despite the extraordinary changes that these initiatives may have brought to the lives and livelihoods of millions of poor people. This book is a collaborative effort to dig out 20 years of experience of a successful Nepali project that sought to give the most remote and vulnerable families in the country a decent life. After two decades, it was time to take stock of the efforts, processes and results of the project in order to share them beyond the steep boundaries of the Himalayas.

This book was written within the framework of the Leasehold Forestry and Livestock Programme as part of an initiative of the Government of Nepal, the International Fund for Agricultural Development (IFAD) and the Food and Agriculture Organization (FAO) to promote knowledge management in rural development. The results can be seen at: *www.lflp.gov.np* .

Through an extensive collaboration which brought together the efforts of a team established and coordinated by Benoît Thierry, IFAD country programme manager and Jim Hancock from FAO's Investment Centre, the book gradually came to life. The collective work of drafting and editing, which was led by Brett Shapiro with the support of Govinda Kafley, Pashupati Koirala and Jaap Reijmerink, was also supported "on the ground" by Bala Ram Adhikari, Project Manager, Nav Raj Baral, Consultant, and for editing by Kenichi Shono from FAO Bangkok.

From the end of 2011, project staff have been encouraged to collect all historic paper and electronic documentation (reports, analyses, databases, posters, photos and videos) in order to create a comprehensive electronic library of the project over the past 20 years. From January 2012 to September 2013, technical notes on project results and impacts were prepared by project staff and the FAO technical assistance team through a knowledge management operation managed by the Investment Centre. These documents also provide an overall picture of the project's 20-year evolution and its various phases. During the same period, articles based on interviews and first-hand accounts of villagers and project specialists were produced or updated (Brett Shapiro, Lorina Sthapit, Kaushal Shreshtra, and IFAD). The articles focused on important changes that took place in the daily lives of the project's beneficiaries. In addition, Daniel Cabral da Silva and Louise Duquesne, from ISTOM tropical agriculture school Paris, prepared case studies based on field interviews, and PROCASUR (an organisation specialised in Knowledge Management conducted several "learning routes" with project beneficiaries.

Finally, Brett Shapiro spent May and June 2013 visiting the villages in the project area to record the villagers' "stories" and to sift through the existing documentation to compile this book. His inspiration and talent helped to shape the extensive information and, chapter after chapter, tell the story of LFLP. He knew how to take the story down two paths: recounting the daily life of the communities, and explaining the more technical aspects of the project. He was accompanied by Naresh Newar, a journalist who produced some of the stories with InterPressService (IPS).

In 2014, the draft was reviewed and rewritten thanks to the contributions of a peer review group from the Government of Nepal, FAO, IFAD and development partners. Jim Hancock and the FAO team conducted an analysis of project results and impacts, and Nicolas Savajol, GIS specialist, created a variety of maps of the project outcome. Yangyuentham Wanaporn revised the draft and created graphs from available data. Kaushal Shreshtra selected the photos and wonderfully formatted the graphic material for the book layout with Simon Bichet who finalized it.

Toward the end of 2014, CABI showed interest in having it published, on the same model that was done some years ago for a book about the Mandrare project in Madagascar entitled *Nourishing the Land, Nourishing the People* (L'Harmattan 2007 and CABI 2010).

The original idea of such books is to capture a resonant experience of rural development and make it accessible to everyone – development practitioners and the curious alike – and not to limit it to the standard administrative document that is ultimately buried under mounds of others.

Thanks to this publication the steep boundaries of Himalaya leasehold forestry groups open and unveil the reality – the process, the challenges, the results – of a long-term development initiative that made a difference.

Benoît Thierry
*October 2014*

# Foreword

Writing this book would not have been pleasurable – let alone possible – without the guidance, support and infinite patience of so many people. Words cannot capture the thanks I wish to extend to Navraj Baral, a kind of guardian angel who spirited me from village to village, organizing each and every visit, serving as indefatigable interpreter during the hours of conversation I had with villagers, answering my never-ending questions that each conversation provoked, and in the rare moments of spare time teaching me about local trees with an enthusiasm that a parent feels for a child. His energy and his passion for his work, not to mention his expertise, were infectious and inspired me to navigate the difficult sections of this book in the same way that he inspired me to navigate the many treacherous footpaths leading to the more remote villages.

Very special thanks also go to Govinda Kafley and Pashupati Koirala, whose wealth of knowledge about leasehold forestry transcends all documentation written about it, and whose meticulous scrutiny of the various drafts of the document ensured that the figures were accurate and supported the narrative, and that the narrative was accurate and supported the figures. So much of what they shared was not to be found in reports, brochures, websites and other written testimony.

I would also like to thank the many project and government staff who spent time with me in the field and in their office explaining their perspectives and sharing their knowledge so that I could stitch together a complex but coherent book that would be forthright and without the "hard sell" position that is implicit in so many books of this genre.

And finally, I must thank the hundreds of villagers who welcomed me into their homes, took precious time away from their labours to sit with me under a tree, walk with me through the leasehold land and talk to me about their life, their needs, their hopes, their aspirations. This book is for them, and the millions of others like them, in the hope that it contributes to effecting positive change in their lives and livelihoods.

Brett Shapiro

*October 2014*

# Introduction

Leasehold forestry is a new kind of property rights regime that started in Nepal about 20 years ago. It has dual objectives: regenerating degraded forest land and alleviating rural poverty. Under this system, the Nepalese government hands over state-owned, virtually open-access, degraded forest lands to a group of the poorest and most vulnerable households, on average ten in number, but ranging from five to fifteen. Each household is eligible to receive around one hectare of land in the form of a group lease contract. The duration of the lease is 40 years, with a provision to extend it another 40 years, covering four generations of the same family. The state requires the households to protect their forest lands against degradation from open grazing, forest fires, soil erosion, etc., either for the purpose of enhancing the natural regeneration of trees, shrubs and grass or for cultivating economically beneficial perennial and multi-purpose plants.

This book seeks to capture the leasehold forestry experience, with a special focus on the Leasehold Forestry and Livestock Programme, which has been a critical part of the leasehold forestry experience in Nepal. It will describe the origins and evolution of the various leasehold forestry programmes and projects that have been launched in the country, as well as their objectives, components and activities, and concrete outcomes and impacts. However, it will also emphasize the story behind the numbers and the individual lives that were impacted by the initiatives. Testimonies are a strong element of the book, as they reveal the on-the-ground results of the leasehold forestry experience: What was successful and why? Why did some things work well and others not? Testimonies come from programme, project and government staff and, most importantly, the farmers themselves.

It is hoped that the information and the stories captured in these pages will help practitioners and policy-makers in Nepal and in all countries affected by the same development issues to understand and draw useful lessons about leasehold forestry, and adapt, mainstream and scale up those elements of leasehold forestry that will promote sustainable livelihoods for poor rural people while preserving the extremely precious natural resource on which their lives and livelihoods depend. In these years of climate change, the programme's impact on livelihood as well on the environment is worth to be counted and disseminated through this book.

**Women farmers are taking the lead in managing leasehold forestry programmes in rural Nepal. Credit: Naresh Newar/IPS**

**JHIRUBAS, Nepal, June** 2013 Nearly 300 km from Nepal's teeming capital, Kathmandu, in a small village dug into the steep slopes of the mountainous Palpa district, 35-year-old Dhanmaya Pata goes about her daily chores in much the same way that her ancestors did centuries ago.

Pata and the roughly 200 other residents in the scenic yet sparse Dharkesingh village, part of the Jhirubas village development committee, live off the surrounding forests, in bright red, thatched-roof mud huts. Jhirubas is the most remote of the 3,913 village development committees scattered across 75 districts in Nepal, but it shares with its counterparts a high level of underdevelopment, food insecurity and poverty.

The road infrastructure is very weak and often gets washed away in the monsoon rains, making transportation of food very difficult – in fact, over half the population suffers from inadequate food consumption. The nearest water source is a three-hour walk away. These villagers have no illusions of living in grand circumstances; their humble dreams consist only of ensuring a decent future for their children. And with the help of a massive leasehold forestry programme, they are doing just that.

In 2005, a 12.7-million-dollar Leasehold Forestry and Livestock Programme (LFLP) took flight in 22 mid-hill districts, stretching from the country's easternmost extremity to its western border, covering 28,000 hectares of forest land managed by nearly 6,000 forestry leasehold groups involving 58,000 households.

Four years later the government began pilot projects – led by the DoF, with technical inputs from the FAO and financial assistance from the Government of Finland – in four districts including Palpa (where Jhirubas is located), where locals have converted degraded forest areas into the country's largest broom grass plantations. Locally known as "amriso", the grass now covers 246 hectares of the 350-hectare region. The grass requires little water and thrives on steep slopes, preventing landslides and helping to remediate the soil.

By turning the flowers of the plant into traditional brooms, which are then sold to a local retailer, villagers earn the money required to stock up on food for the monsoon months when the roads in their landslide-prone village become impassable. "In the last 12 months we earned about 3.5 million rupees (roughly 35,000 dollars) and the income is growing every year", Navindra Thapa Magar, a local farmer and secretary of a leasehold forestry cooperative in the Kauledanda village of the Jhirubas village development committee told IPS.

Each of the 246 households in the village earned about 150 dollars in 2012, income that has proved to be indispensable in supplementing villagers' diets during the nine months out of the year when production of maize, wheat, potatoes, millet and green vegetables comes to a standstill. Amreso leaves also provide fodder for livestock, and the stems provide fuel.

**Women run the show**

Households surviving on less than 80 dollars per year quickly stood out as the target population for the project, which promised each family a 40-year free lease of one hectare of land.

DoF and FAO officials provided support by training farmers and initiating a shift away from slash-and-burn practices, known locally as "khoriya farming", towards more sustainable agro-forestry techniques, in which crops are interspersed with trees and other plants, ensuring a longer and healthier life for the entire ecosystem.

What officials had not anticipated, however, was the level of women's participation in the project. A wave of male migration out of Jhirubas over the last few decades had pushed women into the dual role of labourer-housekeeper. Daman Singh Thapa, chairman of the Kaule leasehold forestry cooperative, explained that when the scheme spread to their remote village, women quickly took up the challenge of planting and harvesting the grass, working long hours on the steep slopes.

Govinda Prasad Kafley, Team Leader, TA-LFLP, an FAO administered project, added that every participating household now involves equal numbers of trained men and women, who share decision-making power. While FAO experts say income generation has led to developments like the installation of water pipes, which relieve women of having to walk several kilometres each day in search of water, others worry that the burden of farming and business operations heaped on top of household chores and care of livestock might end up hurting rather than helping the community.

Forty-year-old Bom Bahadur Thapa told IPS that the work, which includes hand-clearing shrubs in order to plant the grass and then hand-picking the flowers for the brooms, is backbreaking. "Let's hope that men become more involved, instead of leaving to look for work elsewhere", she said.

Indeed, news of the project's success has already gone viral, prompting migrant workers to return to their village after pictures of thriving broom grass plantations and the smiling faces of their families replaced images of hardship.

To reduce the drudgery of harvesting and carrying brooms on their backs to the local collection centre, several farmers in the community recently pooled their resources to purchase a tractor, becoming the first leasehold forestry group in the country to do so.

With the grass providing plenty of fodder, livestock herds have increased four-fold from roughly two to three goats to an average of 12 goats per family, said Hasti Maya Bayambu, chairperson of a leasehold forestry group in Dharkesingh. The community is even considering selling the excess fodder to markets outside their village.

Following the success of broom grass plantations, impoverished families from the traditionally marginalised janjatis (indigenous) and dalit (low caste) groups have also embarked on commercial ventures, producing cardamom and ginger using agro-forestry techniques, according to Palpa District Forest Officer Suresh Singh.

But even while celebrating the project's success, government officials are gearing up for the next big challenge: what to do when aid from the FAO and IFAD expires at the end of 2013, leaving farmers without technical inputs like free seeds, savings schemes and marketing trainings that are integral to the proper functioning of the micro-economy that has developed around the programme.

Narayan Bhattarai, the hub officer and key field officer of the pilot districts, told IPS that farmers rely greatly on the presence of fulltime field officers, who, in addition to arranging trips for officials and donor representatives, boost locals' confidence in the project.

By the farmers' own admission, it will take at least five years to attain full self-sufficiency. Unless donor agencies step up their efforts, the future of one of Nepal's most successful rural development programmes hangs in the balance.

Article by Naresh Newar. Inter Press Service News Agency. This story was captured during the field mission drafting this book in June 2013. It was then published online by IPS: http://www.ipsnews.net/2013/06/leasehold-forestry-brings-a-new-lease-on-life/

# *Chapter 1*

# Rural Nepal: its people, its forests

*"Standing on the beautiful hilltop*
*Looking at the distant snow-capped*
*Mountains hiding behind clouds...*
*I look at the stretched green field*
*Down the hill at the smoking valley*
*And instantly realize that*
*It is not the dream of heaven unknown...*
*But it is the sips of air that I take*
*For the well-being of the world*
*And of the hope of all green..."*

*From Sips of Air, a poem by Dhal Bahadur Jirel, Nepal*

## A brief history of Nepal

It is theorized that the word "Nepal" was derived from the Sanskrit "nipalaya", which means "at the foot of the mountains" or "abode at the foot", a reference to its location in relation to the Himalayas.

***Prehistory:*** There is evidence that people of Kirat ethnicity lived in what is now Nepal more than 2,500 years ago and ruled the area for about 1,225 years (800–300 BCE). Their reign had 29 kings, the first of whom was Elam (also known as Yalambar), who is referenced in the epic Mahabharata.

***1700s – the beginning of modern Nepal:*** Modern Nepal was created in the second half of the 18th century. The king of the small principality of Gorkha, Prithvi Narayan Shah, formed a unified country from a number of independent hill states. At an early age, he dedicated himself to the conquest of the Kathmandu Valley and the creation of a single state, which he achieved in 1768.

***1800s – internal turmoil:*** After Shah's death, his dynasty began to expand the kingdom into what is present-day northern India. At the same time, his heirs were unable to maintain firm political control over Nepal. Rivalry between Nepal and the British East India Company over the states bordering Nepal and India led to the Anglo-Nepalese War (1814–1816). In the end, large parts of the Nepali territories of Terai (nearly one third of the country) were ceded to the British in exchange for Nepalese autonomy. These territories remained in India when India became independent in 1947.

The Rana dynasty ruled the Kingdom of Nepal from 1846 to 1953. The Shah monarch was reduced to a figurehead, and Jung Bahadur was the first ruler. He codified laws and modernized the state's bureaucracy, and in 1855 he attempted to impose his influence in Tibet but was stopped in the Nepalese–Tibetan War (1855–1856). In 1885 a coup d'état took place. The nephews of Jung Bahadur killed his sons and took control. The Rana dynasty developed into a power family and is still very influential in the country. The family formed close alliances with the Shah dynasty through marriage and business.

### *1900s – from "monarchy" to "democracy"*

**1950s.** Popular dissatisfaction was growing among those who were educated in Indian schools and also among the Ranas themselves – many of whom were marginalized within the hierarchy. They wanted to liberate Nepal from autocratic Rana occupation. After an armed revolt, the Shah family returns to power and appoints a non-Rana as prime minister. A period of quasi-constitutional rule follows.

**1959.** A new constitution was issued, and the first democratic elections for a national assembly were held. However, years of wrangling between kings and the government ensued, and in 1960 the democratic experiment was dissolved.

**1960.** King Mahendra carried out a royal coup and promulgated another new constitution. Members of Parliament and hundreds of activists were arrested in the process. The new constitution established a party-less system of panchayats (councils) and solidified the absolute power of the monarchy, keeping the King as head of state with sole authority over all governmental institutions.

**1979.** King Birenda (King Mahendra's son) called for a national referendum to decide on the nature of Nepal's government: either the continuation of the panchayat system with democratic reforms, or the establishment of a multi-party system. The former system won, and the King carried out the promised reforms.

**1991.** Nepal held its first parliamentary elections in nearly 50 years. The Nepali Congress won 110 of the 205 seats and formed the first elected government in 32 years.

**1996.** The Communist (Maoist) Party of Nepal began an insurgency in five districts, establishing a provisional "people's government" at the district level in several locations. The rebellion escalated.

**2001.** The royal family was assassinated and soon after, the new king temporarily deposed the government and took complete control.

**2005.** The King declared a state of emergency to quash the revolution, placing politicians under arrest, cutting telephone and internet lines, and curtailing freedom of the press.

**2006.** Strikes and street protests forced the King to reinstate the parliament. A seven-party coalition took control of the government and stripped the King of most of his powers.

**2008.** In the elections, the Maoists secured a largest-party status, with the prospect of forming a government to rule the proposed "Republic of Nepal". The newly elected Assembly declared Nepal a Federal Democratic Republic, thus abolishing the 240-year-old monarchy. Ram Baran Yadav (Nepali Congress) became the first president, and Pushpa Kamal Dahal (Unified Communist/Maoist Party) became the first elected prime minister.

**2013.** November, new elections were conducted after Parliament failed to finalize and promulgate the new Constitution.

## The country at a glance

Landlocked between India and China, Nepal has a total area of 147,181 square kilometres. Altitudes range from less than 100 metres above sea level to more than 8,800 metres in less than 100 km distance. With this variety of altitude and corresponding climates, the country is rich in biodiversity, and has a huge potential for niche agricultural products. Natural resources are abundant, and various agro-climatic conditions are favourable to developing food and cash crops. However, the physical isolation and rugged terrain of Nepal's hilly and mountainous regions make it difficult to promote economic activities and deliver services.

About half the population lives in the Plains (Terai), the tiny southern belt of Nepal; 43 per cent live in the Hill areas (up to 2,500 metres); and 7 per cent in the mountain areas that are more than 3,000 metres above sea level.

In recent years, political instability has been another major obstacle to development efforts. Poverty, lack of economic growth, and increasing marginalization contributed to political unrest and violence. A Maoist rebellion that began in 1996 in the remote hill districts of the Mid-Western region later intensified and spread across large parts of the country. More than 14,000 Nepalese were killed during the major conflict period and about 600,000 were internally displaced or made homeless. In addition, more than 2 million people are believed to have fled to India. Fighting occurred largely in rural districts, starting in the West, taking a heavy toll on agriculture. In November 2006, a comprehensive peace accord was signed between the then Royal Government and the Maoists, which marked the ending of the armed conflict. The accord also included clauses about the transformation of the country to a multi-party democratic republic as well as about respect of human, social and economic rights. A reinstated parliament abolished the monarchy, formed an interim government led by the Nepali Congress leader and provided an interim constitution for the nation. A Constituent Assembly election was held in April 2008, and the country entered into the era of the "3Rs" – reconciliation, rehabilitation and reconstruction. Although the armed conflict has ended, peace and the political situation remain fragile. Following the peace accord, Nepal has been making the arduous transition from a decade-long conflict to a democratic society. However, political uncertainty persists, with no agreement reached over the key issues of governance structure.

Despite these obstacles, Nepal has achieved notable gains in reducing poverty: the "incidence of poverty" (i.e. the share of the population whose income or consumption is below the poverty line) decreased from 42 per cent in 1996 to 33 per cent in 2006, to 25 per cent in 2010[1], and 23.25 per cent in 2012[2]. In addition to the expansion of road and communication networks in the country (in 2013, almost each of the 75 districts had road access), these gains have come about mainly through a significant increase in remittances resulting from the mass migration of adults and youth in search of alternative livelihoods from villages to cities, neighbouring countries and abroad. With about 1,000 men travelling abroad for work every day, remittances now constitute a significant proportion of Nepal's gross domestic product (GDP) – 23 per cent in 2011 – and they keep growing. At the same time, more women are heading households alone and taking on the burden of sustaining the rural economy, a phenomenon called the "feminization of agriculture". Women constitute more than 60 percent of the agricultural labour force but have little access to land, production technologies and training.

There is progress and there are obstacles. When taken together, however, Nepal is one of the poorest and least developed countries in Asia and in the world. Its per capita GDP is about USD 630 (World Bank, 2011), and nearly a quarter of the total

---

[1] National Living Standards Survey NLSS III.

[2] Central Bureau of Statistics, 2013.

population lives below the national poverty line. There are serious problems of food insecurity and malnutrition throughout the country, especially in the hilly and mountainous regions where 50 per cent of children are malnourished. This is largely due to the poor performance of the agricultural sector (ref. 19 + IFAD website).

### Lens on the rural areas

About 80 per cent of Nepal's population lives in rural areas and depends on small, fragmented subsistence farming for their livelihoods. Household food security and poor nutrition are still widespread in rural areas. Most households have little or no access to basic social services such as primary health care, education, clean drinking water and sanitation services. They have generally very small landholdings and a few are landless, have high rates of illiteracy and are also concentrated in specific ethnic, caste and minority groups, particularly those of the lowest caste (*dalits*, or untouchables) and indigenous peoples (*janajatis*). Life is a constant struggle for survival. The most vulnerable and marginalized groups are the lowest social castes, indigenous peoples and women.

Poor rural people in Nepal include:

- destitute people, such as sick or disabled people, abandoned children and displaced persons;
- extremely poor people, including illiterate or landless people or those with very few assets;
- moderately poor people, such as those who have small farms that produce insufficient food for family consumption or for income generation, and thus are often heavily indebted;
- people who are "nearly poor", including small farmers who are at risk of slipping deeper into poverty as a result of factors such as conflict, debt and land degradation.

Land ownership – especially of productive arable land – in Nepal has traditionally been concentrated in the hands of a few. For most poor rural families, access to land is extremely limited. Almost

70 per cent of households have holdings of less than 1 hectare and many of them depend on plots that are too small to meet their subsistence requirements. The growing population has put huge pressure on cultivable land, especially in the Terai, which also supports many landless migrants from the hills. The growing population also has led historically to unsustainable use of natural resources, including overgrazing and deforestation in hill and mountain areas. And erosion in the uplands causes flooding in the lowlands that can be devastating to crop yields. Social discrimination in rural Nepal plays a significant role in keeping the most disadvantaged people poor and marginalized. Excluded groups include smallholder farmers, landless labourers, lower castes, indigenous peoples and women. Discrimination on the grounds of caste is officially illegal in Nepal but is in fact widespread, especially in rural areas. Members of the lowest caste (*dalits*) are the most disadvantaged group, and many of them work as wage labourers for higher-caste and well-off farmers.

**Figure 1: Distribution of caste and ethnic groups in Nepal**

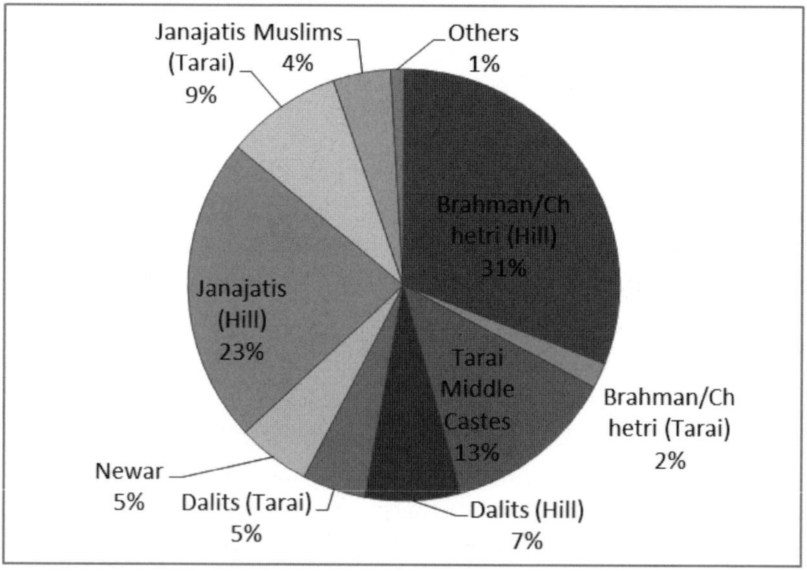


Source: Friedrich Huebler, May 2007, huebler.blogspot.com

There is a wide gap between women and men when it comes to access to health, nutrition, education and participation in decision-making. Infant mortality is much higher for girls, and illiteracy is far more common among women than men. Many rural women live in severe poverty, without any means of improving conditions for themselves and their families. Within households women often have less to eat than men. Insufficient calorie intake can lead to chronic malnutrition in the infants they feed. Poor families are often obliged to send their children to work rather than to school. In this way the poverty cycle is perpetuated in the next generation. It is estimated that about one quarter of the children in Nepal are engaged in some kind of family or wage labour.

**The voices of women**

*Look at my grass here, it's very close to the house. Before, I had to walk four or five hours to collect it. We also produce vegetables now, which are good for the children. And with fodder so close, we can feed the animals in stalls, and the children can study instead of herding livestock.*

*Sanu Babu Udas*

*Before I had leasehold land, for about six months a year I used to leave the house from 6 a.m. until 4 p.m. collecting ground grasses. From July to October, ground grasses are available in my small field, and from October to January they are now available in my leasehold forest plot. It takes me just three hours at most, which saves six to seven hours a day for other activities.*

*Goma Danuwar*

Women in the hills of Nepal spend enormous amounts of time and energy each day collecting water and fodder or fuel wood, tending to their children and taking care of their domestic, livestock and farming activities. Additional tasks fall on them when the men migrate in search of employment. The biggest change noted by the women was the time it takes for them to conduct their daily tasks.

Now they are able to accomplish more and different tasks, largely because of the time they save collecting grass, fodder and firewood, which are more plentiful, closer to their homes and located in places with which the women are familiar, thus minimizing the need to wander in order to gather these essential items. Many women also highlighted that their time savings have been accompanied by substantial technical assistance, knowledge and credit, which have better equipped them to use their new-found time. Several women, for example, reported that they had used credit to acquire livestock and that the time savings resulting from the leasehold forestry had enabled them to look after the cattle. Secure access to the leasehold forest land and the significant savings in time and work that have resulted have also enabled more children to attend school. Before, children's labour was required for grazing livestock and fetching fodder. In many cases, the leasehold savings groups have contributed towards the costs of sending children to school. Several women highlighted another important factor leading to an increase in school attendance: their husbands had become more receptive to their arguments about the need to send the children to school.

Source: IFAD/Shapiro, B. 2001. Voices from the Field. Women's access to land and other natural resources in Nepal. Woman's Resource Access Programme

## Forest resources in Nepal

Forest is one of the major natural resources in Nepal. According to Nepal's Department of Forest Research and Survey, forest is estimated to cover about 29 per cent of the national territory. In addition, 10.6 per cent of the country is covered by degraded shrub lands.

During the 20th century, these valuable resources have generally been decreasing and Nepal became famous in the 1980s for exporting arable land to India, through massive erosion.

Demographic growth was one of the main causes of degradation, as more and more people exploited forest to cover their needs for energy, fodder, etc. The Ministry of Forest and Soil Conservation estimated in 1988 that 75 per cent of the total energy consumed in the country, as well as 40 per cent of fodder for livestock, was extracted from forests. This was clearly unsustainable.

However, the amount of forest degradation and the reasons for it vary throughout the country. Between 1978 and 1994, forest area decreased at an annual rate of 1.7 per cent throughout the country, whereas in the hills it decreased at a rate of 2.3 per cent annually. Between 1964 and 1991, Nepal lost 570,000 hectares of forest. In the Terai, forest was mainly destroyed to be converted into cultivable land or urban areas; in the hills it was degraded by excessive and improper use, leading to an increase in shrubs areas (ref. 15).

Recent macro-level studies and visual interpretations have revealed that Nepal's forest coverage and condition is significantly improving due to the Community Forestry intervention. For example, in a report on 20 Terai districts published by the Department of Forests in 2005, the deforestation rate had been slowed down from 1.3 per cent to 0.27 per cent outside the protected areas.

Therefore in 2014, almost 40 per cent of the country can be considered as forest land which is on the increase for the past decade and a considerable result of forestry public policies in Nepal.

## Importance of forest for rural people's livelihoods

Forest is one of the main assets of Nepal's economy. FAO has estimated that Nepal's forestry sector contributed 3.5 per cent to the GDP of the country from 1990 to 2000. This figure does not include the many indirect effects of forest, in terms of soil conservation, livestock production, water resource control and tourism. The Ministry of Forests and Soil Conservations puts the figure at 15 per cent.

For poor rural people, forest is the cornerstone of their livelihoods because it provides fodder for livestock, stabilizes the soil and provides suitable agricultural land under its cover (for non-timber forest products, which generate cash incomes and provide food for self-consumption).

Given the importance of forest for the livelihood of Nepal's rural population, it was urgent for the Government of Nepal to halt forest degradation by implementing new models of forest management that would be ecologically sustainable without compromising the lives and livelihoods of those who depend on forest.

## Community forestry emerges

Nepal's forestry policies have evolved considerably during the past 50 years. In 1957, all the forest areas were nationalized, and restrictions on tree cutting were adopted in order to lower the very high deforestation rate that the country was facing. Previously, the elite had access to and ownership of vast tracts of forests in the hills, numbering in the thousands of hectares and spread throughout the country. When nationalization was introduced, the elites were no longer interested in protecting the forests, which were subsequently transformed into a no-man's land. This was the main cause of degradation at the time.

In 1961, a new law extended these measures to all the lands surrounding forests that had been under fallow for at least two years. But due to corruption and illegal cutting, the measures were not sufficient to stop deforestation; the situation became even worse in some areas.

In 1978, considering the failure of the nationalization system, a new forestry policy was adopted, giving responsibility for forest exploitation and protection to "panchayats". Panchayats were local administrative bodies, gathering representatives from many villages and usually representing between 2,000 and 4,000 people. Panchayat Forest and Panchayat Protected Forest Rules led to the creation of the first official Community Forestry Development Project in 1980. The project was funded by the World Bank, with technical assistance from the Food and Agriculture Organization of the United Nations

(FAO) (ref. 15). Thus, if the nationalization system had not been imposed (with its failures), community and leasehold forestry would not have been conceived.

Community forestry actually originated and evolved through the National Forestry Plan, 1976, which, for the first time, saw a need to hand over the nation's forests to the local level. The recommended policy then was to hand over the resource to the village council, a politico-administrative unit. In 1977, the policy received legislative back-up through an amendment to the Forest Act (1961). The programme eventually received a more official commitment when it was endorsed by the National Planning Commission in its 6th Plan.

A decade of trials showed that the village council was not a proper unit for such handover. Instead, the local people with an indigenous form of use rights could be a better alternative to whom the concerned resource might be transferred. This very experience actually formed the basis for the Master Plan for Forestry Sector policy in 1989, which envisaged handing over all accessible forests in the hills to communities of user groups "to the extent that they are able and willing to manage them". The policy received legal backing when the old forestry legislation was replaced by a totally new set of legislations: Forest Act 1993 and Forest Regulations 1995.

Policy priority to community forestry means that the forests cannot be transferred to any other form of tenure arrangements unless no demands exist for community forestry (ref. 25).

**Community Forestry – how it works**

When a community forestry user group (CFUG) is formed, anyone from the community is eligible to be integrated in it. The District Forest Office (DFO) has the authority to hand over National Forest to a group of local people.

There is no legal limitation, neither in terms of surface of the community forest nor in terms of number of beneficiaries in the CFUG, the only limit being the actual capacity of group members to access and exploit the community forest. Boundaries of the forest handed over are identified in a first phase, followed by definition of the rights and responsibilities in a constitution paper. Then, an operational plan is written to define the forestry use policy. This paper must be signed by both CFUG and DFO.

Communities are generally handed over the forest for an undetermined period of time, and are guided by a periodic management or operation plan of five or ten years. As long as the groups work according to the mutually accepted plan, the land cannot be taken back. Renewal is only for the management plan. In those cases when the area is taken back, the DFO is obliged to form another committee and hand over the forest again as community forest.

Source: ref. 15

**CFUG rights as per the Forest Act (1993) and Forest Regulation (1995)**

**1. Right to self-governance**

- Communities have rights to form a Community Forest User Group (CFUG) as per their willingness, capacity, and customary rights.
- Community forest boundaries will not be restricted to existing administrative or political boundaries.
- Government can dismantle the CFUG if the latter is found to engage in large scale deforestation but it is the duty of the government to reconstitute the CFUG.
- CFUGs can elect, select or change executive committee anytime.
- CFUGs can punish members who break their rules.
- CFUGs can amend or revise their constitution any time.

**2. Right to forest management and utilization**

- There is no limit to the forest area that can be handed over to communities.
- CFUGs can make optimal use of their forest by growing cash crops together with forest crops.
- CFUGs can mortgage their standing forest products with financial institutions to obtain loans.
- CFUGs can utilize their funds for any purpose (but 25% of income from forest must be spent in forest development)
- CFUGs can freely fix prices and market their forest produce.
- CFUGs can establish enterprises and make profits.
- CFUGs can seek support from any organization.
- CFUGs can raise funds by various forestry and non-forestry means with all income going to group funds with no requirement for sharing financial revenues with government.
- CFUGs can invest in any areas, persons or development activities according to the decision of CFUG assembly.

Sources: Pokharel et al. (2008): Forest Act 1993; Forest Regulation 1995.

(Note: table taken from ref. 15)

## Leasehold forestry emerges

The origin of the lease concept dates back to the amendment of Forest Act (1961) in 1977. However, the concept was far from being considered until the 1989 amendment to the Act when provisions were made to lease out forest lands to poor families. The Master Plan for Forestry Sector made a provision for a classification of the nation's forests into five broad categories including "leasehold forest". But at that time the concept seems to have been inspired by an idea of leasing out land essentially for commercial purposes. In consonance with the policy stipulations, the Forest Act of 1993 essentially stuck to commercial concepts of the lease and did not make special provision for leasing land to the poor. The Forest Act restricts its objectives to the following stipulations:

- Production of raw materials for forest-based industries.
- Sales, distribution and use of forest products through increased production.
- Forest conservation-based tourism (eco-tourism).
- Agro-forestry with emphasis on forest conservation.
- Bio-diversity conservation.

There was a change in the provision when Forest Rules 1995 explicitly made a provision for leasing forest lands to the poor. Although this provision received second priority to community forestry, it did receive priority over the commercial or industrial lease.

**Forest Act of 1993, Section 30**

"Priority to be given to the Community Forest: Notwithstanding anything contained elsewhere in this Act, any part of the National Forest suitable to handover to the Users' Group as Community Forest shall not be handed over as Leasehold Forest."

In 1998, the National Planning Commission declared leasehold forestry as priority programme for poverty alleviation. This declaration was soon followed by the promulgation of Leasehold Forestry Policy 2002, which indicates that forestry land could be granted to:

- Commercial forestry enterprises.
- Entrepreneurs for eco-tourism.
- Households living under the poverty line.

Leasehold forestry is a new kind of property rights regime with the twin objectives of regenerating degraded forest land and alleviating rural poverty. Currently, around 7,000 leasehold forest user groups (LFUGs) with 65,400 household members function in 40 out of the 75 districts in Nepal.

**Table 1: Coverage details of programmes**

| *S.N.* | *Programme type*** | *Total districts covered* | *Number of groups* | *Area (ha)* | *Beneficiary households* |
|---|---|---|---|---|---|
| 1 | HLFFDP | 10 | 1,768 | 7,421 | 12,221 |
| 2 | LFLP | 22 | 4,101 | 20,450 | 39,465 |
| 3 | TA LFLP | 4 | 81 | 1012 | 925 |
| 4 | WUPAP | 10 | 901 | 11,620 | 12,597 |
| 5 | BISEP-ST | 8 | 26 | 133 | 236 |
| 6 | LFP | 15 | 71 | 60 | 790 |
| Total | | 40* | 6,948 | | |

*Districts overlap ** See Chapter 5.

**Leasehold forestry – how it works**

The government hands over state-owned, virtually open-access, degraded forest lands to a group (LFUG) of poor households, generally less than ten in number, with each household receiving around one hectare of land in the form of a lease contract with the group. Prospective LFUGs undergo a social assessment to determine their eligibility. Only the poor are legally eligible to receive leasehold forestry. The two eligibility criteria are: "owning less than 0.5 hectares of land" and earning at maximum "an annual per capita income of 3 035 NPRs", which corresponds to about USD 110 at 1985/86 prices. For 2010/11 per capita income is NPRs 19,261 per year (USD 192), reflecting the change in living standards. The duration of the lease is 40 years, with provision to extend it to another 40 years. This lease enables the recipient household to exercise legally all rights exercised by a private landowner, although actual ownership of the land is vested with the state. The state requires the groups to protect their forest lands against degradation from open grazing, forest fires, soil erosion, etc., either for the purpose of enhancing the natural regeneration of trees, shrubs and grass or to cultivate economically beneficial perennial plants. Open grazing on the leasehold forestry land is to be replaced by stall feeding of livestock. Since the land is classified by the government as forest land and not agricultural land, forestry staff cannot promote cereals. In addition, leasehold lands are generally steep and therefore cereal cultivation is likely to cause erosion. Leasehold groups are authorized to extract forest products, distribute them among the group members and sell surpluses to outsiders in accordance with provisions made in the operational plan.

Leaseholders are responsible for protecting any surviving old and large trees on the leased land, but these trees remain the property of the government. Leaseholders can transfer or sell their rights to others after they have successfully completed one-third of their lease period. They cannot, however, sell the leased land or easily pledge it as collateral for obtaining loans. The basic idea is to enhance forest regeneration while making it possible for the land to meet basic livelihood needs. The programme expects leasehold forestry households to enhance their income in a sustainable manner from: livestock (mainly goats), due to improved fodder availability; and from planting and selling non-timber forest products.

Rangers from the DFO, with the help of staff from the District Livestock Services Office, help LFUGs to prepare a forest management plan of five or more years. LFUG members are also provided with technical advice and training, in order to help them restore the forest on their plot, and start income-generating activities. The groups are also provided with basic material, such as seeds, in order to reduce investment costs poor people would otherwise not be able to afford. A pair of she-goats and one buck are also provided as a grant. LFUGs have a more democratic system of decision-making than the more commonly elite-dominated traditional communities. Since the groups are small – between 5 and 15 households with an average of 9 members per household – a homogeneous group is formed with no dominating figure. Their rights and responsibilities are stated in their constitution or rules and regulations.

Leasehold forestry is considered innovative for a number of reasons. First, it utilizes degraded forest or wasteland that has low productivity but is widely available: approximately 11 per cent of the total forest area of Nepal is shrub land and therefore appropriate for conversion to leasehold forestry land; in addition there are 500,000 hectares of uncultivated land. Second, the land can be handed over to the resource-poor population for the twin purposes of addressing poverty and regenerating forests.

However, because of the policy priority for community forestry, only a fraction of the potential area is used for leasehold forestry. Since community forestry began in the late 1970s and all the government forest land (including barren land) was handed over before leasehold forestry started, it is difficult to know how much of it actually remains. A process has been introduced in the name of "Land Allocation" to provide land to the poorest people in community forestry as well, Similarly, there have been instances when parts of community forestry that were barren and suitable as leasehold forestry were returned to the government to be handed over as leasehold forests. For example, a recent survey conducted by TA-LFLP in four districts outside of the LFLP area (Gorkha, Palpa, Nawalparasi and Tanahun) estimated that 4,500 hectares of forest was available for leasehold forestry for intensive cultivation.

## Community forestry and leasehold forestry: complementary approaches

Community and leasehold forestry are very comparable in many ways. Both are meant to conserve forest and empower rural people. Community forestry and leasehold forestry are both systems based on transferring the usufruct right of the forest from the state to groups of local people. The idea is that if rural people are provided with forest plots, and are allowed to generate benefits from them, they will manage their plot in a sustainable way, in order to preserve their productive assets. The result is then a win-win situation, as rural people can generate additional income with their plot, and protect forest from degradation – a task that state bodies hadn't been able to accomplish until now.

At the same time, the co-existence of these two systems of forest management has raised discussions over the country, with many advocating one system by criticizing the other, and others advocating the implementation of a forest policy mixing both approaches. As we shall see there is much complementarity, and further opportunities for building synergies. (These issues are taken up in Chapter 6.)

**Table 2: Community and leasehold forestry on the ground**

| *Community forestry* | *Leasehold forestry* |
|---|---|
| Bigger and richer forest land managed by larger communities irrespective of wealth status. Intended for deriving benefits in the medium term. | Small and often degraded forest patches managed by relatively poor people, who by definition fall below the poverty line. Intended for deriving benefits in the short term. |
| The group normally manages forests in totality. Group might decide to allocate forest areas to smaller hamlets close to a certain section of the forest. However, land division between individuals is not a normal feature. | Groups decide on whether and how to allocate land. Some groups manage the leasehold forest collectively, sharing in the benefits on the basis of the work performed by each household. Others allocate individual plots, by lottery or an informed judgment of the productivity of each plot. Individual plot size then reflects differences in productivity. |
| Income-generating activities considered and undertaken. | Emphasis given to income-generating activities through pasture/livestock-related developments. |
| Major thrust is forest management. | Thrust is on livelihoods/income-generating activities. |

| | |
|---|---|
| The group owns the funds in common. These may be used for community development works (schools, community drinking water supply, etc.) but not for private purposes. The benefit accrues to the group and the group decides where and how to use it; it is generally not used for funding household chores, except in the land allocated to "sub groups" where the leasehold forestry concept has been introduced. | If the group has decided to allocate individual plots, generated funds are purely private and individuals may decide how to use the funds (excludes support provided by outside agencies specifically for commissioning development works in the community). Otherwise, funds belong to the group, although even here, individual households can use funds based on their needs. |
| Group membership is dynamic. Those who move from the locality lose their membership and those who migrate into the territory may negotiate for membership. By the same token, the offspring of members automatically inherit the membership if the family splits or after the death of the person having the membership. | The group manages its membership, and inheritance is ensured by the group. Migration is only an issue if the entire family leaves. If only one or two family members leave, then the family remains a member of the group. No family will easily give up its entitlement to 40+ years of user rights, and one issue is that a group cannot easily absorb new members if there is no additional forest area. |

Source: ref. 2

**Table 3: How community and leasehold forestry proponents view each other**

| *Leasehold forestry as seen by community forestry proponents* | *Community forestry as seen by leasehold forestry proponents* |
|---|---|
| Competing for the forest land, when community forestry is actually the priority programme. | Insensitive to the situation of the poor, when land allocation has not been introduced. |
| Programme is too expensive and is run with loan money. Output does not match the input and sustainability is questioned. | The programme overlooks the degraded areas and just concentrates on the better-quality forest lands. |

| | |
|---|---|
| Considers only a small section of the community and the forest, and ignores the wider environment. | Focuses on tree products and is not sensitive to the situation of the poor. |
| Field practice and process (forest management) transparency are lacking. | Field practice and process do not focus on the poor and hence the rich reap more benefits from community forestry intervention. |
| Knowledge of group mobilization is lacking. | The fund is mobilized more for social development than poverty alleviation. |

Source: ref. 2

*Chapter 2*

# An initiative takes root – HLFFDP, the first leasehold forestry project

*Everest. Himalayas. Blinding white against brilliant blue. The Roof of the World. Yes, Nepal is the roof of the world and conjures spectacular images. But under this roof, its 26.6 million citizens carry out the business of daily life, some in the isolated mountainous and hilly areas, others in the plains, and others in the major cities. About 83 per cent of them live in rural areas, and most are engaged in agriculture, often with agroforestry and in close relation to the surrounding forests, the agriculture still dominating the economy and accounting for one third of Nepal's gross domestic product (GDP). However, the share of the government budget allocated to agriculture is only about 3 per cent. Lately, there has been growing interest in agriculture in the country and an acknowledgement of its importance to the economy.*

The concept of leasehold forestry was launched on the ground in 1992 in the Mid-hills districts. Between the end of the 1970s and the mid-1990s, deforestation, land degradation and soil erosion spelled disaster for rural households in the districts, where a large percentage of the population is poor.

As forests disappeared, people – especially women – were forced to spend more time collecting fodder and fuelwood, which in turn led to a drop in agricultural labour supply and production, and decreased food security. Community forestry was introduced as one way to address these issues. However, as time passed, it became clear that the poorest people were being excluded from the initiative. Nepal's government launched the Hills Leasehold Forestry and Forage Development Project (HLFFDP), which was financed by an IFAD loan, as well as a grant from the Royal Netherlands

Government for the Technical Assistance component implemented by FAO. Its goal was to reduce poverty and restore environments in the Middle Hills by offering 40-year leases of small plots of degraded, public forest land exclusively to groups of the poorest rural households. The stronger the group, the better chance it had to continue to maintain and improve the site.

**IFAD in Nepal**

The International Fund for Agricultural Development (IFAD), a specialized agency of the United Nations, was established as an international financial institution in 1977 as one of the major outcomes of the 1974 World Food Conference. The conference was organized in response to the food crises of the early 1970s that primarily affected the Sahelian countries of Africa. It resolved that "an International Fund for Agricultural Development should be established immediately to finance agricultural development projects primarily for food production in the developing countries." One of the most important insights emerging from the conference was that the causes of food insecurity and famine were not so much failures in food production but structural problems relating to poverty, and to the fact that the majority of the developing world's poor populations were concentrated in rural areas[3].

Nepal was one of the first countries to benefit from IFAD loans, beginning in 1978. The 14 projects financed to date have mobilized significant resources. For every dollar lent by IFAD, an additional dollar has been committed by the government and other donors.

IFAD's strategy in Nepal supports the development policies and programmes of the government and other partners, especially in relation to peace-building, reconciliation, reconstruction and economic recovery. IFAD's investments reinforce these efforts by

---

[3] http://www.ifad.org/governance/index_full.htm

addressing the issues of poverty, inequality and social marginalization that have been at the heart of conflict in the country.

IFAD recognizes that building sustainable livelihoods and the rural institutions that can support them – particularly in a fragile political and physical environment – is a long-term effort that requires extended support. Its interventions are designed to:

- stimulate income diversification and productive employment by promoting a range of economic opportunities that can bring equitable benefits both on and off the farm
- unleash investment by poor rural people in market-oriented activities, reducing their vulnerability to climate and other shocks by supporting instruments that can mitigate risk
- strengthen rural institutions, enabling them to deliver effective, accountable and climate-smart services to on- and off-farm producers on an equitable and sustainable basis.

In particular, IFAD continues to direct its resources towards the hill and mountain areas, where poverty levels are high and access to infrastructure, services and markets is extremely limited. For maximum impact, the Nepal country programme targets two main groups: vulnerable farm households with sufficient land to develop on-farm activities as their main source of livelihood; and land-poor households and young people, including migration returnees, who can benefit from developing off-farm microenterprises. (From IFAD website).

**FAO in Nepal**

FAO was established in 1945 as a specialized United Nations agency after 44 governments, meeting in Hot Springs, Virginia, the United States, committed themselves to founding a permanent organization for food and agriculture.

Achieving food security for all is at the heart of FAO's efforts – to make sure people have regular access to enough high-quality food to lead active, healthy lives. FAO's three main goals are: the eradication of hunger, food insecurity and malnutrition; the elimination of poverty and the driving forward of economic and social progress for all; and, the sustainable management and utilization of natural resources, including land, water, air, climate and genetic resources for the benefit of present and future generations.

Nepal became a member FAO on 21 November 1951. Since that time Nepal and FAO have been cooperating to improve the agricultural and rural development of the country. FAO was the first among the different UN agencies to start its office and field-level work in Nepal, focusing initially on agriculture and water resource management.

The achievements of 60 years of cooperation is an opportune moment to reflect upon the successes achieved over those years while at the same time reaffirming the commitment and enthusiasm to maintain the momentum into the future. During these 60 years almost 300 projects have been implemented by the organization, embracing a broad range of programmes related to crop, vegetables, forestry, livestock, fishery, food safety, nutrition, planning, policy, rural development and environment conservation and FAO supported the drafting of many agriculture policies and strategies in the country.

**Working together for leasehold forestry**

IFAD support has been crucial in developing the concept of leasehold forestry, which first took the form of the Leasehold Forestry and Forage Development Project (HLFFDP), implemented in ten mid-hills districts between 1992 and 2002. The government continued the project-initiated activities between 2002 and 2005 (the "bridging phase"), extending coverage to 26 districts. IFAD funding was resumed in 2005 for a Phase II of the initiative – the Leasehold Forestry and Livestock Programme (LFLP) – which was to continue for nine years, ending in 2014.

Under HLFFDP, FAO provided a Technical Assistance (TA) component which was designed for the first five years (1992–1997) with the financial support of the Government of the Netherlands. The TA was extended in 1997 and ended with the completion of HLFFDP in 2002. A second phase, consisting of a Technical Cooperation Programme (TCP) project, was implemented from 2007 to 2009. Since then, FAO has been implementing a "Technical Assistance Project for LFLP" with financial support from the Government of Finland. The TA support provides technical backstopping for effective implementation of LFLP and to pilot new modalities and expand to other districts. (Other donor-supported initiatives making use of the leasehold forestry concept are described in Chapter 5.)

The HLFFDP originally had four implementing agencies: the Department of Forests (DoF); the Department of Livestock Services; the Agricultural Development Bank of Nepal; and the Nepal Agricultural Research Council. The DoF was to identify suitable sites and to supervise forest management; the Department of Livestock Services was responsible for the distribution of livestock, fodder tree saplings and forage seeds. Both departments were to assist in drawing up operational plans and providing appropriate training. The Agricultural Development Bank of Nepal was initially chosen to take the lead role in group formation and in providing credit, but it was replaced by the DoF due to its lack of competence and enthusiasm for the role. Nepal Agricultural Research Council activities focused developing technologies for suitable grasses, legumes and fodder trees. With three implementing agencies working at central, district and field levels, coordination was a key factor (ref. 12).

Indeed, the success of the programme hinged to a great extent on the formation of dynamic and cohesive groups in order for a critical mass to be built.

Participants rehabilitated the land by banning grazing and by stall-feeding their livestock, as well as by cultivating fodder grasses, fruit trees and other plants provided by the project. The leases gave them long-term land tenure and along with it the incentive to regenerate, protect and manage the degraded forest areas they were using, while offering them benefits in terms of improved livelihoods.

**Incomes diversify**

"Before the leasehold forestry programme, we worked as labourers for our livelihoods. We were short of forage and fuelwood. At the age of ten our children had to collect forest products instead of going to school. We heard about this programme through the forest ranger. He told us we could get leasehold land if we asked. We worked as suggested, by planting new species and protecting indigenous species. Because of the project we now have time for other activities. I am raising buffalo now, and conditions for the family have improved greatly now that I am making money by selling buffalo milk."

*Mundra Bahadur Magar, member of the Langali LFUG in Chitwan*

When the project ended in 2003, 7,421 hectares of degraded forest land had been handed over to 12,221 poor rural households, and 1,768 leasehold forestry groups had been formed. Once the leasehold forestry approach was successfully piloted and its impact was proven, the government scaled up the approach with its own resources from the initial ten districts to 26 "priority" districts in the hills of Nepal. The original districts chosen for HLFFDP were from the central and western regions, which were near Kathmandu and easy to reach. The new districts were primarily from the central and western regions, but at least several districts from all regions were included.

"The project was a major success", said IFAD's country programme manager for Nepal at the time. "The most innovative and impressive element is that it was the first project in Nepal with both forestry and livestock components to focus exclusively on helping the poorest of the poor rural people. But there is considerable need to include the poorest community members in development projects and to ensure more equitable distribution of the benefits from forests. And because the leasehold forestry approach has been so successful, other development agencies are now piloting the same targeting approach to reach the poorest of the poor."

The project broke new ground in other areas as well. There was a strong partnership between the Department of Forests (Ministry of Forest and Soil Conservation) and the Department of Livestock Services (Ministry of Agriculture Development), so that productive aspects of forestry and livestock activities were linked. And gender and development training for project staff and participants focused on both men and women. This gender approach was presented as a best practice at the 12th World Forestry Congress held in Quebec in September 2003. The project's impact on poverty was highlighted in a case study presented at the May 2004 conference in Shanghai, China on Scaling-up Poverty Reduction: A Global Learning Process.

A project evaluation mission[4] in 2003 and impact studies[5] found that the project had contributed considerably to improving the livelihoods of rural poor people, especially women, and to improving the condition of degraded forests. The National Planning Commission commissioned an evaluation of the project in 2005 and also found it very successful.

One of the project's most significant results was the increase in available animal feed, which decreased the average time women used to collect forest-based fodder from 3.9 hours per day to 1.4 hours. This additional 2.5 hours per day meant that women had more time to pursue other productive activities, like literacy and vocational

---

[4]http://www.ifad.org/evaluation/public_html/eksyst/doc/prj/region/pi/nepal/nepal

[5] Ohler, F.M.J., 2000. The Impact of Leasehold Forestry on Livelihoods and Environment. HLFFDP, Field Document 3/2000. Summary: http://www.fao.org/docrep/ARTICLE/WFC/XII/0214-C1

training and income-generating activities. As a result, cash incomes among leaseholders rose by 24 per cent between 1996 and 1999, and almost half of the new cash earners were women.

An increase in fodder also made it easier for household members to switch from "free grazing" to "stall feeding". Stall feeding reduces the pressure on forest lands and increases the availability of manure, which in turn helps maintain or improve soil fertility. As a result, food production increased and leasehold households were more food secure: From 1996 to 1999, the average period of food security rose from 7.8 months to 8.4 months – a 16 per cent increase. (In comparison, non-leasehold households saw a 4 per cent decrease in food security over the same period.)

The increased availability of fodder and access to credit also enabled a number of poor households to purchase and keep livestock for the first time. The average number of goats owned by leasehold forestry households increased from 3.9 heads to 4.4 heads over a 3- to 5-year period, not taking into account goats that may have been sold in the interim. Leaseholder households were also encouraged to expand their livestock to include high-producing animals. As a result, more livestock products are now sold and consumed in leaseholder communities, which has led to improved nutrition and food security, as well as increased incomes.

Most leasehold sites have also experienced rapid natural regeneration of herbs and grasses, as well as varying speeds of natural tree regeneration. Average ground cover in new sites was about 32 per cent, but rapidly rose to 50 per cent after one full growing season, and gradually increased to almost full coverage in sites after seven years. Biodiversity also increased. Some leasehold forestry groups developed fruit orchards, and the number of plant species in two sites increased by 57 and 86 per cent between 1994 and 2000.

The evaluation and impact studies showed that after ten years:

- 1,768 LFUGs had been formed, comprising 12,221 households

- annual household incomes rose from USD 270 to USD 405 from leasehold forest sources
- higher incomes translated into greater food security and improved diet for children
- the number of goats increased from an average of 3.9 to 4.4 per household
- the amount of animal feed and forage increased significantly
- women spent 2.5 fewer hours a day collecting forage and firewood
- women's self-esteem and confidence rose because they had more time for income-earning activities and to attend meetings, training and literacy classes
- school attendance increased because there was less need for children to herd grazing animals
- environmental degradation reversed at most sites; ground cover increased from 32 per cent to 50 per cent after one growing season, eventually reaching 100 per cent coverage
- biodiversity increased significantly; in two sites, the number of plant species increased by 57 per cent and 86 per cent.

**Popularity of exotic forage grasses; failure of exotic fodder trees**

Imported forage grasses and legumes proved generally popular and successful, although in some sites there were problems with the increased tree canopy and competition from unpalatable plants. At higher altitudes, the problems were the traditions of open grazing, and the unsuitability of temperate pasture species for the cut-and-carry system. However, the majority of households noted that the supply of forage was now adequate, while two-thirds had found it inadequate before the project. Sites in all project districts reported income of NPRs 2,000–4,000 (USD 20–40) from green forage.

The planting of imported fodder tree species was not a success, as many sites are unsuitable because of insufficient topsoil, steepness of slope, lack of moisture retention, and competition from noxious vegetation. If fodder tree species are to be imported, their selection will need to be highly site-specific. Many farmers believed that in the leasehold sites natural regeneration of forest cover following a ban on grazing was a preferable approach.

Source: ref. 12

Under the second phase of the Technical Assistance component, 46 women Group Promoters were recruited to assist in the formation and support of groups, and NGOs were hired to select and train them. Their role proved to be fundamental to creating and sustaining the groups and in strengthening women's decision-making role.

**Figure 2: Percentage of decision-making role**

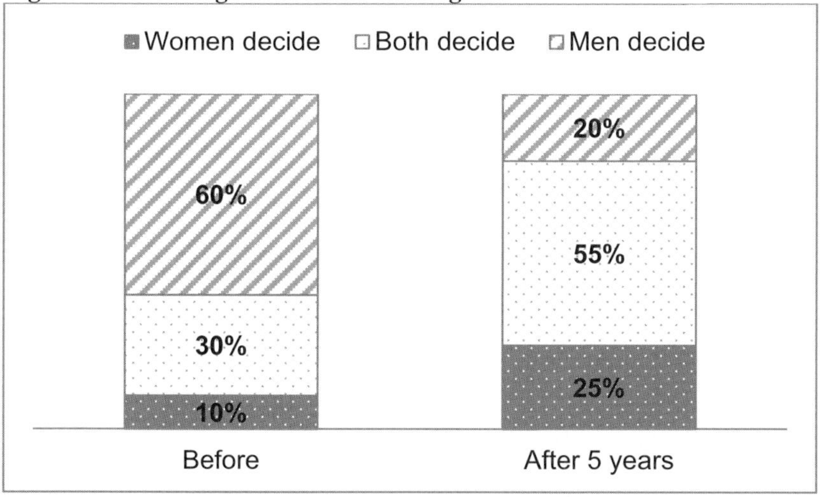


Source: ref. 12

**Tapping human potential – one woman's remarkable transformation**

"When I became a member of the first leasehold forestry user group of my village in 1992, I could barely read or write. I was soon elected secretary of my group because most of the other members weren't even able to write their names. It was hard for me because I was very shy to speak with anyone outside of my family. Only after participating in some trainings I started opening up to the people from my group and I also learned how to read and write very well.

A few years later I received a letter inviting me to go to the District Forest Office in Hetauda and speak about a job as a Group Promoter. I didn't know where the office was or how to get there, but that day a car from the office came and picked me up near my village. At the office two people were waiting for me and speaking Hindi. I was very frightened because I couldn't understand what they were saying, and I had heard stories of girls being kidnapped and trafficking and the like. After a few minutes, they told me in Nepali to follow them and we went to the District Forest Officer. He asked me if I would be interested in working as Group Promoter. I said that I would like to but didn't know what I would have to do and maybe I needed training. Two weeks later, a letter arrived inviting me to go to Kathmandu for training. I was very excited, but I didn't even know where Kathmandu was. An officer and another girl came with me, and after a very long trip on a bus we arrived in the city. We went straight to a hotel where a room had been booked for the two of us. The Officer explained that we could take our meals downstairs and that the next morning he could come to our room to take us to the training. We immediately closed all the windows and the curtains, but we didn't know what electricity was and couldn't find a way to get rid of the light. We couldn't fall asleep for a long time because of the light. We also started fearing again that we had been sold.

In the morning, we heard someone knocking on our door. We were so afraid to open it because we thought they would be clients. The knocking went on and on. Finally I opened it with a sinking feeling in my heart. It was the officer. He escorted us to a bus. We travelled for more than an hour and didn't feel confident at all. When we finally reached the training centre we saw many girls and women bathing and talking, and so many clothes hanging to dry. At that point we were sure we had been sold. I was so sad that I had left my village. When we started talking with other girls, we finally realized that this was actually a training centre!

The training went on for seven days and was so interesting. When it finished we returned home happy and ready to begin our work. At the time there were only 15 leasehold forestry user groups. My main responsibilities were record keeping, credit stimulation and extension of forestry groups in my area.

Since I started working as Group Promoter, the number of groups has increased to 48. I always try to get the women to lead them. Now that the groups are strong and united, I'm also trying to extend them when and where it's possible. Over the years I have taken many more trainings and worked with many governmental and non-governmental organizations involved in the programme. This work really helped me improve my skills and allowed me to link the leasehold forestry user groups to many different agencies and projects. For example, women's fertility, children and adult education, health and children care projects are now starting in our area, and I think many more agencies will provide us with useful trainings in the coming years.

I'm very grateful to my family and to the leasehold forestry programme for what I am now. Everyone knows me in this area for what I do and this makes me so proud. I also feel so much gratitude to my community for listening to me and supporting me during all these years and gave me the platform to work with."

*Ms Suntali Maya Praja, Group Promoter, Raksirang Village, Makwanpur District*

In addition, the emergence of intergroups and cooperatives was a very significant development, which the project had not foreseen. Clearly, social mobilization strengthened links among people, and the leasehold activities made them eager to engage in additional economic activities. The formation of 120 intergroups and 19 multi-purpose cooperatives by 1,600 leaseholder groups paved the way for savings to be mobilized and for marketing initiatives to be developed. Group savings is an essential element in each LFUG, as it serves as the vehicle for providing small loans to members. The existence of these groups also brought in grants to build culverts and bridges, improve trails and footpaths, renovate schools, and complete small drinking-water supply projects.

Perhaps most important were the clear gains the group members made in terms of confidence and self-esteem. They were far more ready to engage officials (and visiting missions) in discussions about their needs and problems. They were also able to exchange and learn from each other's experiences, and to build the critical mass necessary to have a degree of bargaining power. In this process of group formation, the concept of cooperatives began to take root: during the project, 19 cooperatives were formed. The concept of cooperatives would be formalized in the LFLP.

Indeed, the success of a programme site hinged to a great extent on the dynamism and cohesion of these groups. Since in the HLFFDP leasehold forest user groups usually consisted of ten or fewer households and are less educated and more socially disadvantaged than the community as a whole, the challenge remains to enable them to form and maintain functional organizations that last the duration of the lease. A number of years after the close of the HLFFDP, during the LFLP implementation, FAO conducted a Group Categorization Survey, in which the groups were given points according to their status in institutional, forestry, livestock and microfinance aspects and categorized into Active, Medium and Passive groups. The survey analysed 5000 groups and found that even groups that were 15 years old or older, long after project support had been completed, were still very active, especially in forestry activities. Individual members interviewed during missions were adamant about never relinquishing their user rights which they obtained from their group. This certainly paves the way for sustainability, and shows how central secure tenure is in rural livelihoods.

LFUGs were supported with a number of basic infrastructure developments such as providing small drinking water stations, building culverts and small footbridges, and maintaining or renovating school buildings. The infrastructure-related achievements included the building of 38 culverts/bridges, the maintenance or renovation of 366 schools, the completion and maintenance of 231 small drinking-water supply projects, the improvement of 35 hectares of terrace, and the upgrading of 514 km of trails and footpaths (ref. 12).

### *Lessons and challenges of first project*

Upon the successful implementation of the project and lessons learned from it, the government institutionalized group leasehold forestry in 2002 with a revised Leasehold Forest Policy. The government ran its programme for three years (2003–2005), and the Planning Commission of Nepal recommended that the project be extended to 26 districts (covering all the development regions) within the government's own budget.

From the outset, the government had given community forestry priority over leasehold forestry. This meant that there had to be a community consensus before leases for forest land were given to the poor. Since leases were only given for degraded forest, this consensus was usually not difficult to achieve. But after the degraded sites became green and productive, those who did not benefit from leasehold forestry could feel resentment towards those who did. Conflicts could arise, and leaseholder groups tend to be in weaker positions since they are poor.

The outcome of such conflicts will depend upon the decision-making process in handing over community and leasehold forest. Ideally, the local community should agree upon which part of the forest will be deemed "community" forest, which part will be allocated to the poor, and which households qualify as "poor".

Other factors came into play as well and are still relevant. For example:

- the selection of sites in very marginal mountain areas where regrowth is extremely slow and the sowing of introduced grasses so labour-intensive that the resulting income scarcely justifies the efforts;

- the lack of understanding and training in some sites about optimum livestock capacities (the TA for the later project conducted a study to determine the optimal herd size of goats based on available resources);
- decreased productivity of *stylo* grass caused by the failure to re-sow after five years and the increasing tree canopy and replacement by other type of fodder;
- the lack of continuing support for groups in remote locations; and
- the failure to genuinely to respect farmers' preferences and indigenous knowledge in the preparation of operational plans.

The project was a costly one, with an average expenditure of approximately USD 800 per household or USD 1,400 per hectare of degraded land. Can this cost be reduced?

Can and should leasehold forestry be integrated with community forestry schemes?

Can leasehold forestry be replicated by means of existing government personnel, skills, equipment and infrastructure?

Some of these themes and questions were to be returned to in subsequent projects (ref. 12).

*Chapter 3*

# The initiative gathers force – LFLP, the second leasehold forestry project

> "A project is a time capsule. It needs to change into a programme if it's going to be sustainable."
>
> *Mr. Bal Krishna Khanal, first coordinator of the Hills Leasehold Forestry and Forage Development Project*

*From their houses in the hills, slowly, the village men and women converge. Many appear to be timid, somewhat reluctant. Some of the women fall behind, perhaps in deference to the men, or perhaps out of a sense of ambivalence or uncertainty as to whether they should be participating at all. At the same time, they know that they have a great deal that they would like to talk about and that the others would like to hear. Programme staff and group promoters whom they know and trust have encouraged them to participate. They push on.*

*The setting for the discussions is informal. There are no microphones, no round tables, no flip charts with coloured markers. In fact, the setting is outdoors, in their own crop plots, on the grass and under the trees that they visit and use on a daily basis. Later, the discussions will continue during a walk into the forest areas that have been leased to them as part of this government project, as a way of making more palpable the lives that are being spoken about and the changes that are – or are not – taking place. After several hours of walking, talking and observing, themes begin to emerge. Some of them surface again and again, introduced by one participant and enthusiastically elaborated by another and then another. Others speak with a lone voice, but a voice that is so emphatic that it cannot be discounted. How have these people benefited from their newly gained access to land and forestry?*

## Programme processes and organization

To help Nepal implement the leasehold forestry approach as a national programme, IFAD designed the Leasehold Forestry and Livestock Programme (LFLP) in 2005, upscaling the previous projects at a national scale. Once again, degraded national forest or part of it (forests having less than 20 per cent crown cover, open or barren forest land, failed plantation sites, shifting cultivation plots) are handed over to groups of people living under the poverty line.

The LFLP started in 2005 with a slight change in its partners and modalities. Two implementing partners of the previous project – Nepal Agricultural Research Council and the Agricultural Development Bank of Nepal – were dropped and a separate Rural Finance component was incorporated into the new programme to carry forward the rural financial activities of the first phase and with the Department of Forests (DoF) taking the lead role (ref. 2).

However, as the ongoing IFAD-financed Western Uplands Poverty Alleviation Project (WUPAP – see Chapter 5) included four districts from 26 districts which were part of Hills Leasehold Forestry and Forage Development Project (HLFFDP) bridging phase, the LFLP targeted the remaining 22 districts: Accham, Baitadi, Bhojpur, Chitwan, Dadeldhura, Dhading, Dolakha, Doti, Gorkha, Kavrepalanchok, Khotang, Lamjung, Makwanpur, Okhaldhunga, Panchthar, Pyuthan, Ramechhap, Salyan, Sindhuli, Sindhulpalchok, Tanahu and Therathum. About 5.3 million people live in these 22 districts, of whom 2.55 million, or 48 per cent, live below the poverty line. It was expected that by the end of the programme period, some 44,300 households, more than 200,000 people, would have directly benefited. The overall goal was to reduce poverty in the area by allocating leasehold forestry plots to poor families in order to enable them to increase incomes from forest products and livestock. The project was later further supported by a Finnish-funded Technical Assistance (TA) project implemented by FAO pursuing the effort of the Dutch grant in the first phase, which also covered four further pilot districts – more details later. Specifically the LFLP aimed to:

- improve household forage and tree crop production
- improve household production of livestock, especially goats
- provide access to microfinance institutions
- support the government's capacity to implement leasehold forestry in a gender-sensitive way. (ref. 11)

The programme activities were organized into four components, with each component having a number of sub-components, as follows:

**Table 4: Summary of programme components/sub-components**

| *Leasehold forestry and group formation* | *Livestock development* | *Rural financial services* | *Programme management and coordination* |
|---|---|---|---|
| District planning & coordination<br><br>Group formation & forest allocation<br><br>Social mobilization<br><br>Land and forest development<br><br>Forestry implementation support | Goat production development<br><br>Livestock training and services<br><br>Livestock implementation support<br><br>Forage and fodder farming on the farm land | Group savings and credit activities<br><br>Village finance association activities<br><br>Cooperative support | Management of the overall programme<br><br>Coordination of programme activities at the regional and national levels<br><br>Coordination of programme activities with those of other forest management programmes<br><br>Monitoring and evaluation of programme implementation |

Source: ref. 11

The programme builds on the success of the HLFFDP in helping set up leasehold forest user groups that will eventually form leasehold cooperatives or finance associations. Local women are hired as Group Promoters to mobilize the leasehold groups and train them in group management and rural finance. All of the Group Promoters are women.

**"The pivotal role of the Group Promoter"**

Sahana Sona Tamang, a 25-year-old mother of two, is one of them. Since becoming a social mobilizer five years ago, she has saved enough money to send her children to school. Currently, she works with about 280 households to manage their savings and credit, organize trainings and help them share knowledge about what works and what doesn't.

"We are trying right now to find a way to sell our forest products directly to the markets in Kathmandu so we can cut out the middle man and make more money", she explained. "And we meet to talk about how to improve irrigation and manage the unpredictability of the rain."

Despite these and other challenges, Tamang said, people who lease the land have been able to take control not only of the forests, but of their lives. She believes that the future of Nepal's forests rests in the hands of those who live closest to them.

"We are neighbours to the forests here and so it is our responsibility", she said. "Like good neighbours, we help each other."

(ref. IFAD website:
http://www.ruralpovertyportal.org/country/voice/tags/nepal/nepal_forest)

Sona Tamang, 25, a social mobilizer in the leasehold forestry programme in Kavre district, Nepal. ©IFAD/Katie Taft

## A word about the technical assistance provided

Projects such as the LFLP, with the challenging objectives of addressing poverty reduction and natural resources management, and forging new approaches, require considerable knowledge support, and as under HLFFDP, a support TA project emerged. The main objective of the TA was to support the DoF and Department of Livestock Services to improve the effectiveness of the LFLP and build up appropriate institutional and technical capacities at field, district and central levels in order to help institutionalize and scale up of leasehold forestry in the country. The four-year project was supported by the Government of Finland; FAO was designated as executing agency, and the DoF as the lead counterpart agency responsible for project implementation. In four additional pilot

Districts (Palpa, Nawalparasi, Syangja and Gulmi) it also experimented with new landscape approaches, and relationship between community forestry and leasehold forestry.

## Technical assistance: expected outputs

The TA had the following expected outputs. Actual results achieved are discussed in Chapter 4.

**Output 1:** Leasehold forestry guidelines that are linked to poverty alleviation and environmental conservation developed under LFLP activities to be further consolidated and refined following testing in the field.

**Output 2:** A detailed project implementation plan and institutional structure for implementation of the project developed.

**Output 3:** Support and promotion of conducive policy and legal framework for community-based leasehold forestry.

**Output 4:** Effective and efficient human resource development strategies for LFLP developed and implemented.

**Output 5:** Gender and indigenous communities' issues in the leasehold forestry approach strengthened and mainstreamed.

**Output 6:** Livestock and forage production further improved and integrated with ongoing LFLP.

**Output 7:** Strengthened institutional capacity of Rural Finance Institutions and market-based linkages developed under the LFLP.

**Output 8:** Improved Monitoring and Evaluation capacity in DoF and LFLP to effectively manage for development results.

**Output 9:** Development of knowledge management strategy for LFLP to more effectively disseminate lessons learned on leasehold forestry.

**Output 10:** R&D programme to support LFLP strengthened and implemented.

**Output 11:** LFLP piloted in one cluster in five new districts and District Forestry Sector Plan prepared in five programme districts.

## Power in numbers – forming groups

Group formation remains central to the leasehold forestry process: the strength of groups is critical in the manifold activities of the group, like planning, dealing with conflicts, generating savings and getting support from various agencies.

**From the words of one villager**

"In my village the first [leasehold forestry users] group was formed around 1994. We elected a chairman. He was richer than most of us, and after some time he left and migrated to the Terai. We held new elections and I was chosen, I think because I was young and everybody felt I would never leave my village. At that time, the forest had very few trees and the soil was very dry. The District Forest Officer provided us with new seeds and new seedlings and arranged for different trainings. We learned new ways to work our land and to plant trees in the forest. We worked very hard to plant fruit trees, broom grass and forage plants to feed our animals. Before that, we only had a few shifting cultivation plots where we grew beans and maize. We would have enough food for five to six months a year and would have to buy the rest at high prices from the nearest market. It takes almost an hour to get there by foot.

After the programme started we were able to harvest more vegetables, fruit and forage than ever before. We now have enough food for eight to nine months every year. When we started harvesting our new forage, the programme gave us 10,000 Rupees (USD 100) to buy goats. This way we had around 1,000 Rupees each (USD 10). Every family now has between 20 and 25 goats per household, and all of us can even sell a few every year. We also have a lot of broom grass and forage that we sell, and each family earns around 60,000 Rupees per year (USD 600). It's now easier for us to buy the food we need when our own stock is finished. We also attended bee-keeping training. After the training we received bee hives and started producing and selling honey. We make between 10,000 and 15,000 Rupees depending on the year (USD 100 to 150). I'm very

# CHAPTER ONE

## Rural Nepal: its people, its forests

NEPAL AT A GLANCE

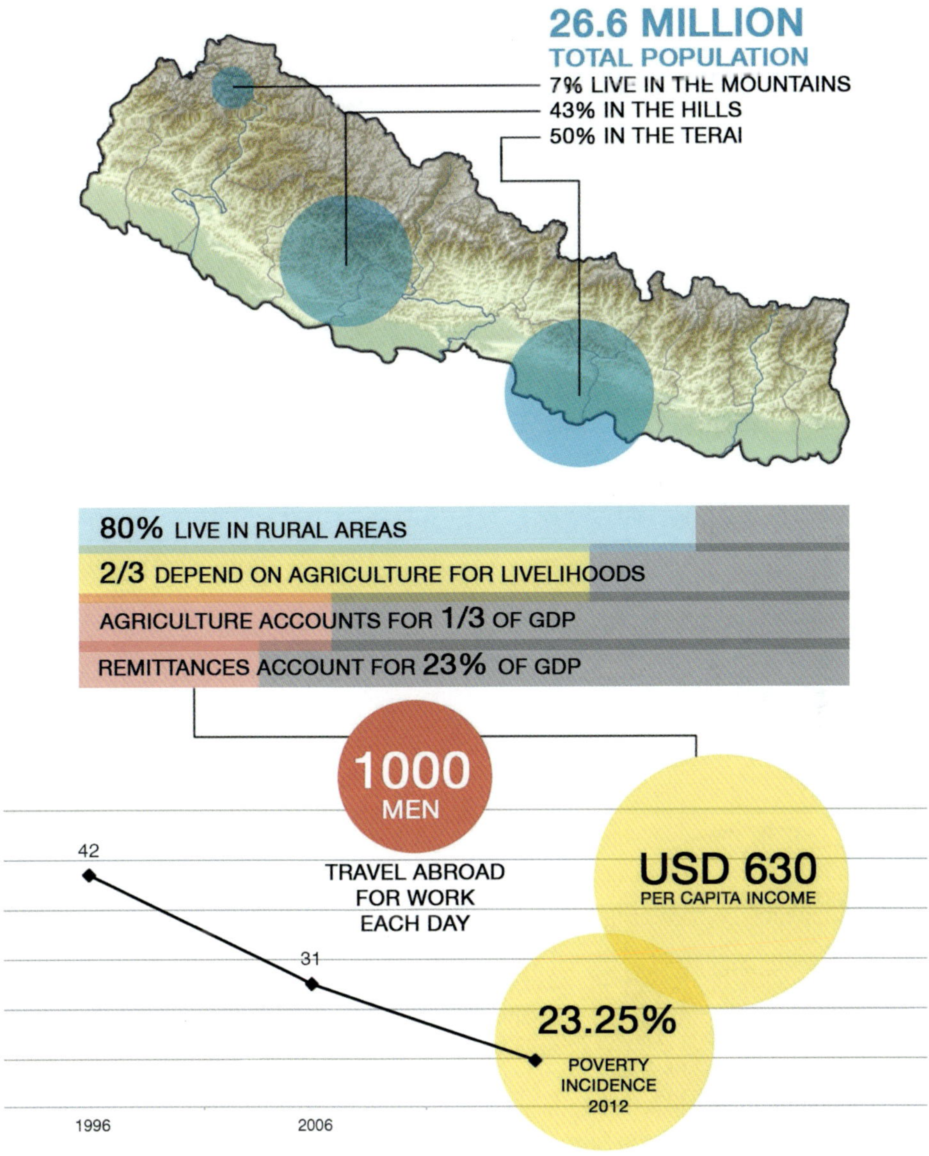


**1.** Nepal is a rural country with a high migration rate.

**2.** The physical isolation and rugged terrain of Nepal's hilly and mountainous regions make it difficult to promote economic activities and deliver development services.

**3.** Devitaar, Kavre district. With her daughter perched behind, a woman plucks fresh spinach for dinner. Smallholder farmers in rural areas have little access to agricultural equipment, financing, or social and economic infrastructures, limiting production systems to a subsistence scale.

**4.** For poor rural people, forest is the cornerstone of their livelihood. A member of Chisapani Puchaar Leasehold Forest Group returning home after collecting firewood

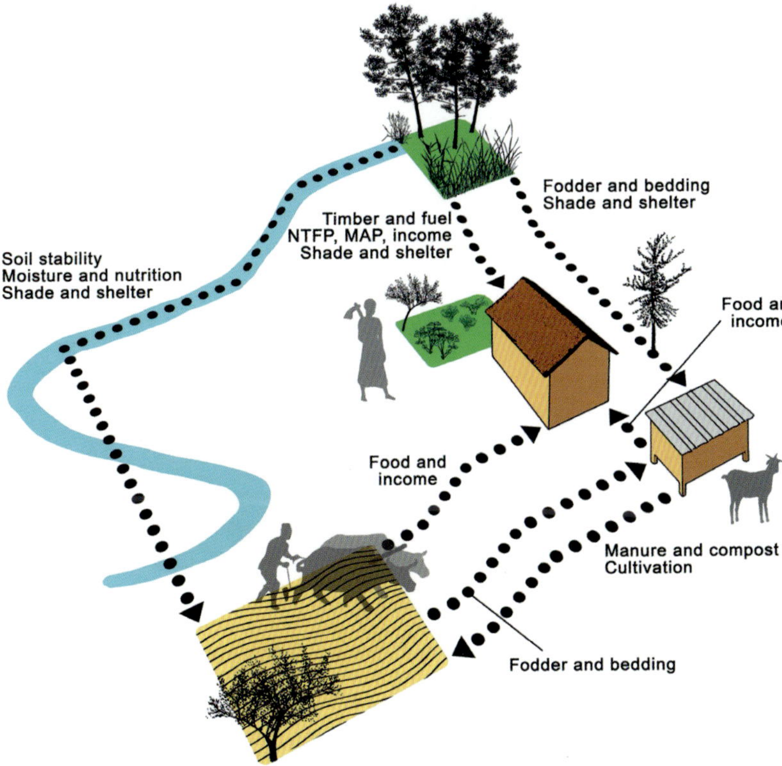


**5.** The LFLP forest–crop–livestock farming system in Nepal is a proven model for rural poverty alleviation.

# CHAPTER TWO

## An initiative takes root – the first leasehold forestry project

HILLS LEASEHOLD FORESTRY AND
FORAGE DEVELOPMENT PROJECT

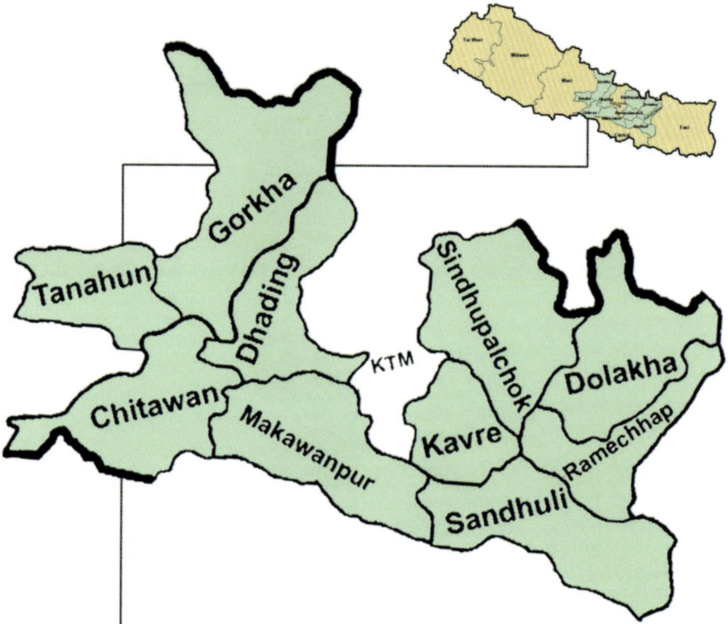

**1,773** LEASEHOLD FORESTRY GROUPS FORMED

**7,457** HECTARES OF DEGRADED FOREST HANDED OVER FOR IMPROVED MANAGEMENT

**6.** This first experience of leasehold forestry dates back to the 1990s and generated promising outcomes.

**7.** Landslide: poor land and forest management practices coupled with steep topography make the hilly regions of Nepal highly prone to soil erosion and landslides, resulting in severe loss of resources.

**8.** Managed LF hillside. Through HLFFDP, 7,457 ha of degraded forests were handed over to poor households for improved management, thereby protecting natural resources and promoting biodiversity.

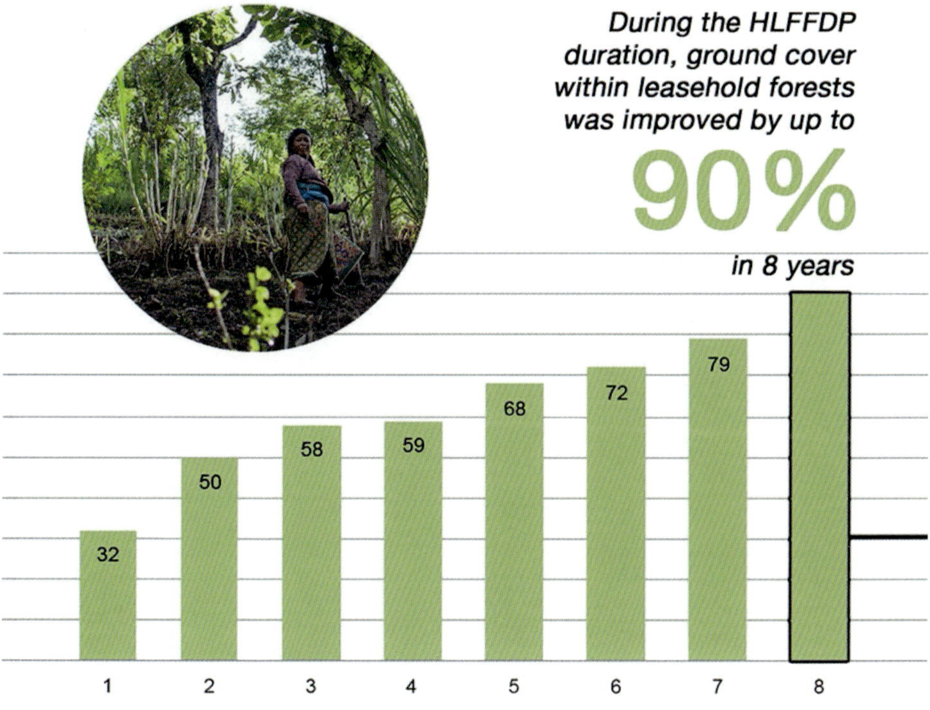


**9.** Users weeding operation Hupsekot Nawalparasi.

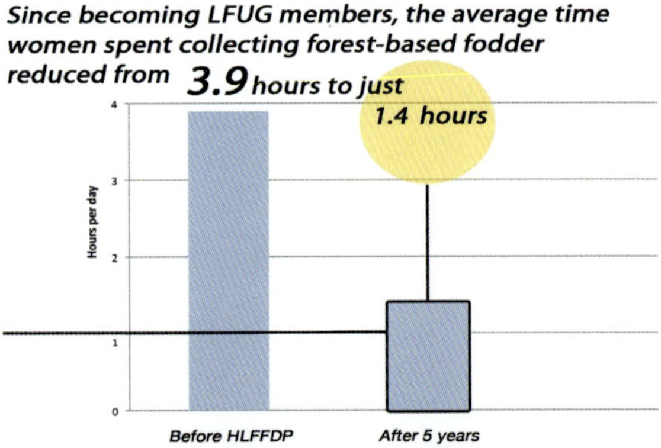

**10.** Women carrying cardamom plants for plantation in the lease land Riyale Kavrepalanchok district.

**11.** LFUG discussion with many women. Gender and social inclusion was an integral component of HLFFDP, significantly contributing to women's participation in decision making.

# CHAPTER THREE

## The initiative gathers force – the second leasehold forestry project

LEASEHOLD FORESTRY AND LIVESTOCK PROGRAMME

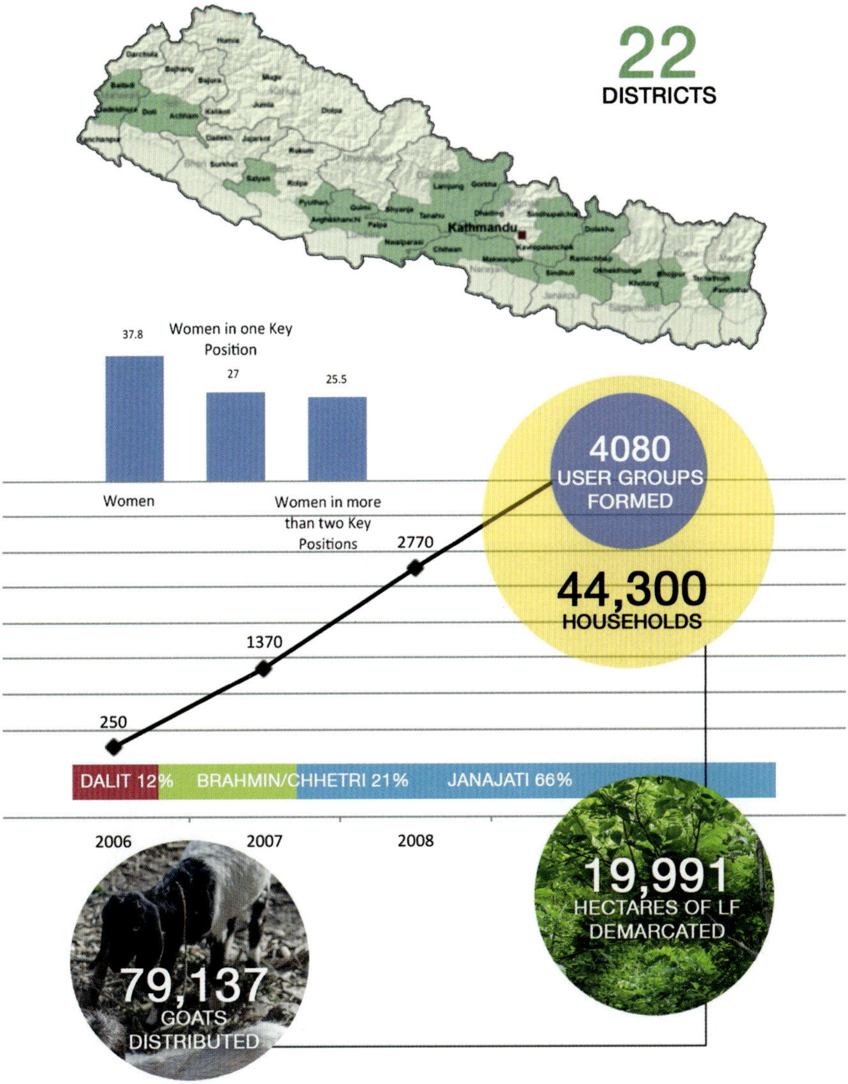


**12.** The scaling up of leasehold forestry at national level became effective with LFLP.

# Broom brings boom: how one remote Nepali village has found the road to harmony and prosperity in Jhirubas

**13.** Grass is gold: brooms being readied for market in Jhirubas.

**14.** Villagers have converted degraded pastures into plantations of grass used for commercial broom-making.

**15.** Barren slopes have been converted into plantations for broom grass.

**16.** Flex FAO poster: throughout the project FAO delivered technical assistance to the leasehold forestry and livestock projects.

NATURAL RESOURCE MANAGEMENT IN LFLP

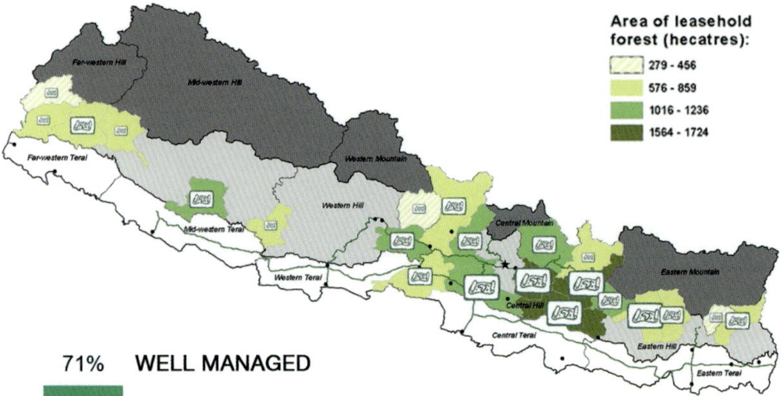


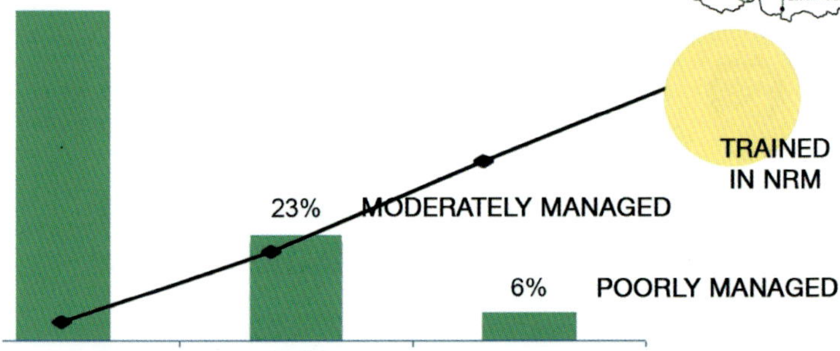

**17.** Tejpat Nursery Mityal and the LFLP impact on forestry.

**18.** Poster Jhirubas for dissemination of new forage plantation techniques.

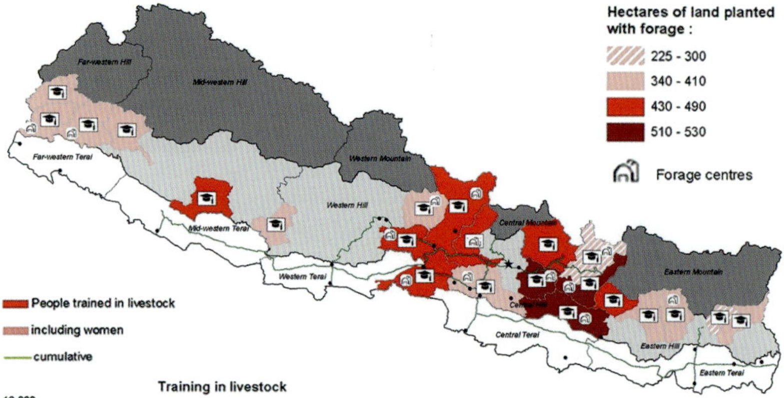

**559**
JTAS TRAINED

**198**
VAHW

**19.** Ram Piyari and her younger son, Bibek, with the goats they received as part of the Leasehold Forestry and Livestock Programme to improve household income in Devitaar, Kavre.

# RURAL FINANCE

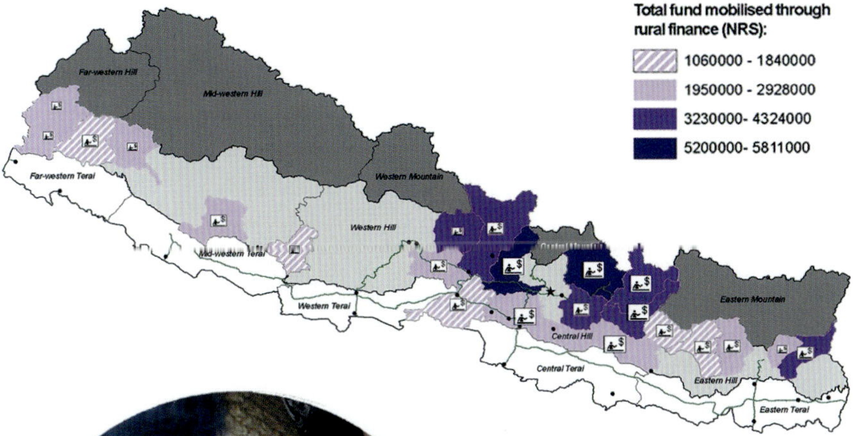

TRAININGS IN GROUP AND FUND MOBILISATION

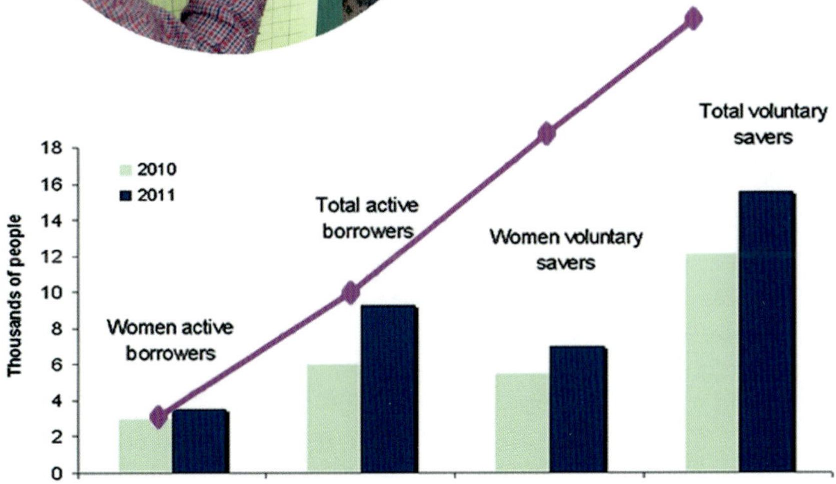


**20.** Beyond forestry and livestock, LFLP supported economic development through provision of credits with self-help groups.

# Cooperatives and LFUGs in Kavrepalanchok District
## - a typical story -

**21.** Chaturman Rayale LFUG Learning Route.

**22.** The dairy farmers working.

**23.** Cardamom producer in his garden.

**24.** Sapkota calculates his income to participants of Learning Route.

**25.** Bhandari - standing, and other participants of the Learning Route.

**26.** Simple testing of fat in milk in the dairy

good at bee keeping and can sell bees to people from other villages, earning around 150,000 every year (USD 150). With the support of the programme, our lives are now easier and we have time to take care of our nursery where we grow seedlings, including asparagus, with the seeds that the programme provided us. We have even saved some money, which can be used to send our children to school. Our village is a more pleasant place, and cooler because the surrounding forest has grown so well."

*Sover Singh Praja Lisidamar, 35, chairperson of one of the leasehold forestry user groups in Raksirang, Hetauda District*

## Identifying the land, forming the groups

Once the District Forest Officer (DFO) delineates the leasehold forest area, group formation is initiated for between five and fifteen poor households. The land is generally demarcated using cement pillars and along alignment hedge plants, such as Jatropha, that serve to conserve the forest areas as well as provide income-generating products. The lease prepared for the groups is initially for a period of 40 years, renewable once. An agreement is drawn up between the DFO and the group. The lease certificate is signed by the DFO under the Department of Forests, and the names of the households of the group members appear in the certificate. The group then prepares an operational plan and a constitution for the leasehold forest in a participatory manner among its members. The chairperson of the LFUG submits the operational plan for the approval of the DFO. The DFO then prepares a lease commitment paper, which the chairperson of the leasehold group signs. The authority to approve the lease contract has been delegated to the DFO since 2006 (ref. 2).

***Selecting the beneficiaries***

To select the beneficiaries, a survey is conducted to determine how many of the community members fall below poverty line. This is followed by a well-being ranking process to determine the poor, poorer and poorest among the eligible poor members. The available land is assessed against the total eligible candidates. If adequate land is available for all eligible candidates (maximum one hectare per household), all candidates are included. But generally, less potential land is found. The community then decides how many are to be accommodated, giving priority to the poorest.

To help groups prepare their operational plans, trained staff of District Forest Officers and District Livestock Services Officers support them. The operational plans must include a Livelihood Improvement Plan, for which the project, with support from FAO technical assistance developed guidelines to be distributed to leasehold forest user groups (LFUGs) and project staff. The guidelines include land use planning, resource mobilization, gender and equity, and environmental issues. As of October 2012, 2,000 copies of the guidelines had been distributed, and 23 trainings were given to project staff at all levels. As a result of the trainings, almost 3,000 Livelihood Improvement Plans were prepared in the districts.

As of December 2012, a total of 6,957 LFUGs had been formed since the first project was implemented. These groups held over 41,000 hectares of degraded forest benefitting about 65,400 rural poor households (LFLP 2013). The average size of forest is 5.93 hectares per group and 0.6 hectares per household (ref. 6).

Conflicts are not uncommon when the leasehold land is being identified and allocated. Before it is handed over, conflicts can arise over boundary claims between private and leasehold land, membership of the leasehold group, and the claims of better-off and poorer families. After the land has been handed over, the main sources of conflict are grazing rights and social issues. Leased land is a limited resource, and when local people see the benefits of leasehold forest, many non-leaseholding households want to join leasehold groups. Such conflicts are usually resolved by local community consensus, mediation from forestry rangers, and the formation of additional leasehold groups where there is high potential for leasehold forestry (ref. 26).

## Developing operational plans

Without and operational plan and constitution, the land cannot be handed over. Therefore, the first step after the group is formed is to prepare the operational plan, which is prepared by the groups themselves with the help of mid-level technicians from the Departments of Forest and Livestock. During and after this step, Group Promoters are on hand to raise awareness and provide other post-formation support.

**Becoming a group promoter**

"When our group was forming, there were few candidates for the group promoter. One criterion was that the person had to be a married woman. I was married. My husband is working in Saudi Arabia. He's been there for two years. Ten men in our area have migrated for work, mostly to the Middle East.

I was so happy to be selected. But I didn't know what I was supposed to do and how I was supposed to do it. But very quickly I received a training on my role as a promoter. And I also received regular guidance afterward. I felt more and more confident.

I received training in many things: how to mobilize savings and credit schemes, how to organize and manage groups, how to promote better health and sanitation, and a special 15-day health service training. Sometimes the trainings were held in other district offices and sometimes even in Kathmandu. It depended on the subject. For the livestock training, I had to travel 500 kilometres away!

After the trainings, I would come home and share my learnings with the groups. I really enjoy my role. The only difficulty for me is when I have to travel alone to remote areas. Sometimes I am afraid. One achievement that I am proud of is that there is no male domination in our groups. There was much discussion, and men and women reached an agreement that women and men will make decisions jointly. Earlier the men used to dominate, but now the men fully respect the decisions made by the women. But when it comes to

> decisions about how income is going to be spent, that is always left in the hands of the grandmother or grandfather. It is about seniority, not about gender."
>
> *Ms Pratrikshya Gurung, Group Promoter, Aamdanda village (Tanahu District)*

The operational plan provides rules of "do and don'ts", a broad framework for developing a detailed plan and accompanying activities, a monitoring system of the LFUG and a Livelihood Improvement Plan. Details of the annual plan and monitoring system are worked out in advance at monthly meetings, in which mid-level technicians from the District Forest Office, District Livestock Services Office and NGOs also participate. The group's General Assembly then approves the plan.

When an operational plan is being prepared or renewed, a ranger (a mid-level forestry technician)/assistant forest officer prepares an inventory of the forest stock in each block or compartment and over the whole leasehold forest area. This inventory provides the basis for planning activities in the forest. The range post (the lowest-level functionary in forestry administration) supervises forest planning at the *ilaka* level: an *ilaka* is a territorial forest office under the District Forest Office, headed by an assistant forest officer and administered by four range posts. (For administrative purposes in the forestry sector, a district is divided into one to three *ilaka* and has eight to 15 range posts.) The *ilaka*-level plan is presented at the district planning workshop and subsequently at the regional planning workshop. (Note that the organization has since changed: Under the District Forest Office, there are now Sector Forest Offices and, under them, Area Forest Offices. The former holds the senior forest officers, and the latter the assistant forest officers, which are the equivalent of the range posts.)

The LFLP integrates forestry, livestock and microfinance organizations. Forestry-related and micro-finance-related components of the annual programme are compiled at the

Department of Forests, and livestock components at the Department of Livestock Services. The departments then forward the programmes to their respective ministries, and they are finally approved by the National Planning Commission. The Ministry of Finance is responsible for allocating the budget, and the consolidated annual programme budget of all sectors is tabled in Parliament for approval in the form of the Appropriation Bill (ref. 26).

**Operational plans, reviews and decision-making – key figures**

On average, more than 70 per cent of the LFUGs are found to be implementing their annual plans and programme in line with their constitutions and forest operational plans. Of the total of 4,101 LFUGs under LFLP, 23 per cent have also developed linkages with other agencies for technical and financial support, while 47.9 per cent of them have been partially successful in bringing on similar support from other agencies. The remaining 29 per cent of the LFUGs have remained passive in implementing their annual plans and programmes as provisioned on their forest operational plans. When it comes to reviewing their work performance, 25.8 per cent regularly review their progress, 51.4 per cent review it occasionally, and 22.8 per cent do not review it at all.

In addition, 25.4 per cent of the groups implement all of their decisions, 45.2 per cent implement at least half of them, and 29.4 implement less than half of their decisions (ref. 7).

### *Group composition and inclusiveness – issues of gender and caste*

The guidelines for implementing the Leasehold Forestry and Livestoc Programme have a provision to include the name of both male and female members of each household as household head. However, in many LFUG constitutions, this provision is not carried over, and only the name of the lead member is mentioned there. About 37.8 per cent of the group members are women. Members are from all kinds of ethnic groups such as *Dalits*, *Janajatis* and other more advantaged members of society. *Janajatis* represent the highest percentage of LFUG members (58.8 per cent) whereas *Dalits* represent only 12.5 per cent. About 21 per cent of LFUG members are *Janajati* women, and about 6 per cent are *Dalit* women. The overall representation of women members in the groups is 37.8 per cent (ref. 7, including figure).

It must be remembered that the main factor of social hierarchy in Nepali society is the Hindu caste system, which reinforces social exclusion among different castes. Even if the system was legally abolished through the Civil Code Act in 1963, consciousness of caste or ethnicity is still strong in most areas where leasehold forestry is implemented. This system is particularly strong in rural societies of the country. Higher castes (such as Brahmin and *Chhetries*) benefit from a dominant social status over the lower castes (such as *Dalits*), providing them with higher political and economic power.

Ethnicity is also a factor of discrimination, which is linked to the caste system, as indigenous people are considered as belonging to lower social layers than upper Hindu castes. In Nepal, indigenous people are often integrated in the caste hierarchy system under the name of *Janjatis*. However, not all *Janjatis* are considered equal, and the social level of a specific ethnic group can vary from one region to another.

Women are also very often underprivileged, in terms of access to education, healthcare and economic opportunities. Being a woman, especially in rural Nepal, comes with the various social prejudices and institutionalized expectations such as having to do the household chores (and serve their husband), to work at a reduced wage, and limited opportunities to participate in community activities.

Social discrimination based on those considerations is a major issue faced in most rural areas of the country, including forest areas, and can make a community approach difficult to implement if community groups are experiencing too wide social and gender inequities (ref. 15).

**The voices of women**

*Previously, I could not talk to men freely, especially outsiders such as government and NGO development workers. I was even scared to talk to my husband's family. Now, however, I have taken training courses and attended local meetings and workshops outside the community. I am accustomed to talking to men and strangers and am happy to do it. I can face men and can even disagree with them.*

*Til Maya Shrestha*

Women's self-esteem and confidence are influenced by many things, including cultural norms and expectations, education, state-ascribed roles, exposure to the outside world, cash-earning ability, decision-making roles and their own inner sense of identity, autonomy and strength. In a part of the world, such as Nepal, where landholding is strongly associated with status and esteem, giving land to poor rural households and, in particular, poor women provides more than just a sense of security; it confers status and prestige unlike any other factor in the society.

Most of the women had not been exposed to development activities before the project began. Community participation has been fostered by the project through the early formation of groups, which are encouraged to hold regular meetings. These meetings provide a forum to discuss project-related issues and general community matters. There are both women's group meetings and mixed meetings. Women's participation is generally greater at the meetings of women-only leasehold forestry groups than at the mixed groups, and it is easier to assure their participation in the women-only groups. A typical meeting lasts one or two hours, and in most cases eight to ten people participate, of whom typically one to three are men. In the all-women group meetings, extremely sensitive issues such as domestic violence are easily addressed. In this regard, many women see the group meetings as both a “protective court” – where instances of domestic violence can be brought out into the open and challenged – and a place to confront social issues and become stronger. The critical building blocks for social mobilization and solidarity are formed in these meetings and pave the way for enhanced visibility and decision-making capabilities on the part of women in community and household matters.

Through the training programme, most of the women have now acquired basic literacy skills, and the group members are much more aware of their legal rights and the importance of education and adequate health, sanitation and nutrition for themselves and their families. Many of the women said that group work makes the task lighter and that their work in the fields and forests is completed in less time. With respect to other leasehold groups, they feel competitive rather than jealous, which strengthens them as individuals and as a group. The feeling of competition has raised their awareness of the need for group solidarity, the appropriate division of labour and the sharing of benefits as well as problems.

Source: IFAD/Shapiro, B. 2001. Voices from the Field. Women’s access to land and other natural resources in Nepal. Woman’s Resource Access Programme

## Who holds key positions?

The LFUGs under LFLP have shown strong inclusion of women and ethnic groups in leadership roles. (It must be noted that in certain regions of Nepal, there are not many ethnic communities, and thus the percentage of ethnic members holding key positions will appear low, although it will correspond to their actual numbers in the group.)

The LFUG chairperson, secretary and treasurer are regarded as key positions in the group. Participation of women in key positions develops their leadership and helps to address their issues and concerns. Women are in two or more key positions in 27 per cent of LFUGs. They are in only one key position in 25.5 per cent of

LFUGs. There are no women in key positions in 47.6 per cent of LFUGs (ref. 7).

*Janajatis* form important ethnic hill tribes in Nepal, and *Dalits* are the untouchable caste, who live in hill and terai communities. *Janajatis* are in two or more key elected positions in 55.1 per cent of LFUGs. They are in only one key position in 10.4 per cent of LFUGs. There are no *Janajatis* in key positions in 34.5 per cent of LFUGs. *Dalits* are in two or more key positions in 21.8 per cent of LFUGs. They are in only one key position in 6 per cent of LFUGs. There are no *Dalits* in key positions in 72.2 per cent of LFUGs, i.e. in most groups. Nevertheless this reflects relatively closely the proportion to their membership, and they are thus largely representative in leadership. We shall see in the impact chapter that overall LFUGs are positively inclusive of marginalized groups.

## *Group cohesion and transparency*

**Speaking out**

Before the leasehold forestry project came along, we used to cover our faces when we saw a new person in our village. We always felt shy and reluctant to speak to outsiders. But slowly and gradually the project staff gained our confidence, and I started opening myself and participating more and more in the meetings. My two siblings felt the same way, and now they also talk freely and frankly. We understand the word "empowerment" and we want more of it.

*Ms Bhadra Kumari Chepang, a young member of a leasehold forestry users group, Godhitar village (Mugling/Chitwan District)*

The LFUGs are supposed to hold monthly meetings to discuss any issues encountered, review the progress of the decisions made at the previous meeting and prepare a plan for the next month. This is one of the obligatory activities defined by their constitutions because it helps the group to remain active and grow into strong organizations, and promotes harmony and group cohesion among its members.

According to a 2012 survey, a total of 63.4 per cent of the LFUGs conducted monthly meetings, while the percentage of LFUGs not holding even a single meeting in the last three months or six months was 14.2 per cent and 25.4 per cent, respectively. Similarly, the percentage of LFUGs that implement the decisions taken at their meetings is not encouraging: LFUGs implementing 50 per cent of their decisions stands at 45.2 per cent; and those implementing less than half of their decisions stands at 20.4 per cent.

Record keeping within the groups is important for transparency as well as to understand how, where and to what extent the group is functioning. Training was provided in record keeping for savings and credit, land development and livestock development. In some groups, all the records of saving & credit, land development and livestock development were well maintained, while in others only the meeting minutes and saving and credit records were kept. Of the 5,042 LFUGs studied, 22.2 per cent kept all the records of saving & credit, land development and livestock development activities; 18.3 per cent kept records of saving & credit and land development or livestock development activities; and 59.5 per cent had only the records of saving and credit. While far from perfect there is considerable dedication to keeping their organizations functioning.

According to a number of studies, group cohesion of LFUGs is weakening somewhat, and the number of LFUGs managing leasehold forestry individually is on the rise – and more prevalent in the groups formed after 2002. A total of 37.3 per cent of groups formed after 2002 reported that they were managing lease land on an individual basis, compared to 27.7 per cent of group members formed before 2002. However, the group management approach is more prevalent in groups headed by women (74.3 per cent) as compared to mixed groups (56.4 per cent) and groups headed by men (64.7 per cent)[6].

**The voices of women**

*Going to the temple and worshipping gods is not enough – going to the community, working and mobilizing the community is what is required.*
*Goma Danuwar*

The importance of women's self-confidence and self-esteem in terms of community development cannot be overstated. Once the women felt comfortable speaking with men within the community, and with those representing the Government and NGOs providing services, they were able to express their needs and hopes. The sense of hope and optimism that comes to women that experience this type of empowerment acts as a catalyst for change and produces individuals that are willing to take responsibility for their own development, despite the additional work that the development activities may require. Such women serve as powerful role models for their children and other women, who desire to imitate them once they are convinced of the benefits of such changed attitudes and behaviours.

Food security is a problem in many areas of Nepal, at least on a seasonal basis: many households are without enough food for three to four months per year. Their only recourse is to seek out moneylenders, who not only charge extremely high interest rates, but also often require borrowers to work free on their land when requested and to give "gifts" such as chickens and goats. Many villagers expressed their satisfaction at the increased amounts of food and livestock fodder that resulted from their access to leasehold forest land. The time they saved by using this land enabled them to initiate other income-earning activities at home, in their small fields near their houses, or by selling their labour. They were relieved that

[6] DVN/NPC, 2005.

their "moneylender days" were over as a result of their access to leasehold forest land and to credit.

Source: IFAD/Shapiro, B. 2001. Voices from the Field. Women's access to land and other natural resources in Nepal. Woman's Resource Access Programme

### *Training as the key*

Training has always been conceived as a major activity in both phases of leasehold forestry. More than 20 different training packages have been provided to LFUG members and government staff, on topics ranging from land development, nursery establishment, seedling production and plantation, forest management, animal husbandry, and seed production to gender, social inclusion and conflict management. Similarly, workshops on planning, coordination and monitoring and evaluation and exposure visits were also organized for the groups and field staff to deepen their understanding of leasehold forestry planning and implementation processes, and to give them hands-on knowledge.

During the LFLP alone, a total of 2,453 training packages were delivered, as well as almost 100 workshops and study tours (ref. 8).

In essence, the breakthrough in social status of leasehold forestry households has been significant. They are socially empowered, aware of government policies and plans, and are more sensitive toward their health and diet. They have realized the value and strength of organizing into groups, and many who used to live in distress are now improving their livelihoods and multiplying their income sources. Most importantly, their voices are now heard by government agencies, development partners, political elders and the community as they form strong organizations. The households have gained confidence and self-esteem, especially with the emergence of inter-groups, federations and cooperatives, which is reflected in the radically different attitudes towards officialdom, and in particular in the much greater readiness among very poor people to engage officials. These achievements can be attributed in great part to the programme's continuous, intensive and targeted social empowerment, group management and leadership coaching, and training programmes.

"More than three dozen research articles, academic studies, and donor and government reports have strongly claimed that LFUGs are emerging as strong pro-poor organizations."

*Navraj Baral, National Expert, Nepal*

## *Collaborating with other agencies*

LFUGs have limited resources. It is not possible for them to implement all planned activities from their own resources. Therefore collaboration with other agencies is essential to pool resources. The study shows that 12.5 per cent of LFUGs have collaboration with more than two agencies, 39.9 per cent have collaboration with fewer than three agencies and 47.6 per cent have no collaboration with other agencies. In general, whatever agencies are available in the village, or in the district, will provide some kind of support. They may include international NGOs, NGOs, line agencies of the government, Village Development Committees, banks, other projects, and the like (ref. 7, including figure below).

**Figure 3: Collaboration with other agencies**

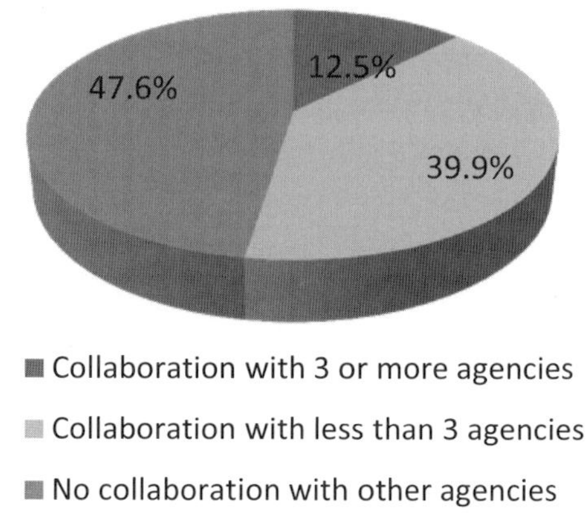


**Power in numbers**

"I will tell you my story of 18 years ago. At that time there was no group unity. We didn't know what a group was. No one thought about helping others. We women were very reluctant to express ourselves and our feelings. And we were reluctant to talk to outsiders. We didn't want to share our thoughts. Our men prepared wine and beer and sold it. Whatever we earned from that money, our men drank up. They misbehaved. We were illiterate. Our health was also very poor because of bad hygiene practices. We followed traditional systems of keeping animals, and didn't know or care much about their health or their feed. Our source of income was firewood and smuggling timber to the local dealers. Afterward, the forest was denuded and the land was barren. When the land was denuded, we started shifting cultivation. That was the situation. That is when the chief of the district forestry office came here.

He told us that if groups of up to seven people were formed, the land would be handed over to us for 40 years and that we would be given support to cultivate it. We weren't terrifically happy, but we accepted it and didn't initially take up much interest in developing the land. At that time, there was open grazing. We thought, why should we plant anything if the animals will eat it? We wanted to drop the idea of leasehold forestry and land management, but when the forest became 100 per cent community forest, free grazing was prohibited. From that we learned to stop free grazing. Norms and regulations started to be put in place about when and where firewood could be collected, and we learned that we should start the same thing in our leasehold forestry. It was then that we really started to appreciate the initiative.

After we formed the groups, we developed our plans and the forest was handed over to us. And also they gave us training on planting and cultivation techniques. We planted ficus species, mango, lychee, jackfruit and banana and pineapple. We also started planting on our farmland. We were very happy to see the forest develop, especially when we started having a surplus of fodder. We then started thinking about livestock. And that is when we really started generating income."

*Ms Prem Kumari Ranamagar, LFUG chairperson, Chitwan District, site of the first leasehold forestry initiative*

In addition to the LFUGs, federations (at cluster and district levels) are an important evolution in the "power in numbers" approach, and their creation has been piloted successfully. All of the groups in one district are federating to form a district-level federation, which acts as a fulcrum for advocacy, awareness-raising, problem solving and negotiating, innovation, leadership training, coordinating with other institutions and, most importantly, creating solidarity among the LFUGs.

## Regenerating the forest and its resources

During the first project period (1992–2002), the majority of leasehold plots handed over were denuded. The prospect of transforming these brown, rocky hills into lush green slopes was a daunting one. Moreover, there were the competing demands of "long-term benefits" – regenerating the forest by planting tree species – and "short-term benefits" – providing sources of income by cultivating forage, fodder and non-timber forest products (NTFPs).

**Out of the shadows**

"In the beginning we were so scared of government people. We didn't want to meet with them. This was especially the situation with women. They would run away when they saw people coming from the outside. The government people had many discussions with us, and we kept asking ourselves should we join or should we not join? It was all so mysterious. We discussed everything and decided that we must join. But we were scattered and not organized. Every household was working in its own way. Members of different government departments arrived and explained to us how to work together. They also invited us to participate in training. Slowly we started to see the benefits and to understand that this project was for the poor.

After the trainings, we planted seedlings. Our plots were completely barren and filled with stones. But after the land development trainings, we adopted what we learned and now it is all covered, and we are self-sufficient in firewood and forests. Before that, our children and women had to travel 4 kilometres from the village to gather wood, and our major source of income was selling firewood which we cut illicitly. We were always afraid of the government discovering us. We thought of ourselves as intruders. After the development of leasehold forestry, our children go to the forest to collect grasses and firewood, and they are back within an hour. At nine they can have breakfast and be at school by ten.

We were so inspired by the training. Those who taught us the skills and technologies were so important."

*Chairperson of one of the LFUGs, Darechowk village (Mugling/Chitwan District)*

To rehabilitate the land, the Leasehold Forestry and Livestock Programme set up 566 nurseries, which eventually produced 29 million multi-purpose tree seedlings. The objective of group nurseries is for group members to raise the seedlings of their own preferred species for planting in order to regenerate and rehabilitate the leased land. Those participating in the nurseries received nursery management training along with seeds and seedlings.

By the end of the period, assessments revealed that the majority of the leasehold land had developed into forests: 71.6 per cent of the leasehold land had been well managed – meaning that forage species covered more than half the area, the number of fodder and other trees increased, at least three species of forage, fodder, NTFP, firewood and timber were present; 23.2 per cent had been moderately managed (forage species covered one third to half the area, the number of fodder and other tree species remained same, and only two species of forage, fodder, NTFP, firewood and timber were present); and 5.2 per cent had been poorly managed – forage species covered less than one third of the area, the number of fodder and other tree species decreased, and only one species of forage, fodder, NTFP, firewood and timber was present.

Forage, fodder and firewood are the major benefits that LFUGs are getting from leasehold forests. Other benefits include NTFPs and timber. Availability of forage and fodder has increased from most of the leasehold forests. The result of the FAO study shows that forage and fodder are now available for more than six months in 13.4 per cent leasehold forests. In 45 per cent of leasehold forests these forest products are available for three to six months. Forage and fodder are available for less than three months in 41.6 per cent of leasehold forests.

**Figure 4: Availability of forage, fodder from leasehold forests**

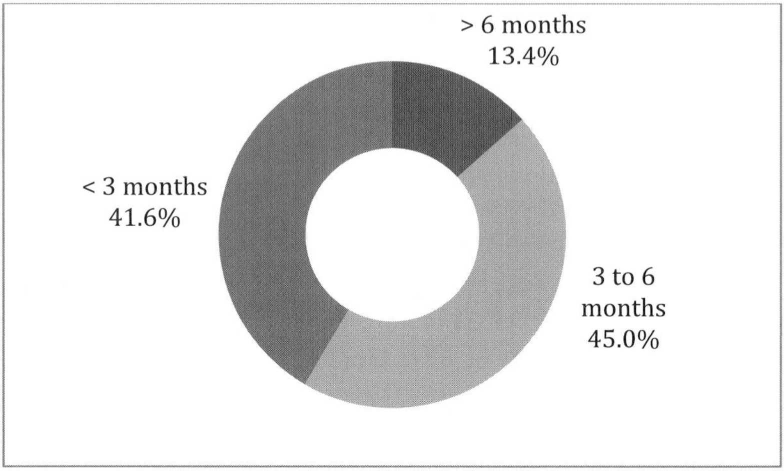


Clearly the most dramatic impact of the programme is its contribution to the rehabilitation of degraded forests. The majority of leasehold plots at the time of handover were denuded and open grazing land, with scattered crown coverage of up to 20 per cent. Now they have taken the shape of natural forests enriched with multiple uses. The rehabilitation has been steady, since leasehold forestry was first initiated. For example, in 2000, forest cover increased from 32 per cent in new plots to 78 per cent in six- to seven-year-old plots, and with a much higher proportion of fodder and forage species: from 30 species in new plots to more than 95 in old plots (ref. 22). In 2005, the National Planning Commission reported that more than 90 per cent of the plots were covered with improved varieties of grass/forage. By 2012, more than 70 per cent of leasehold forestry plots of much larger project area (14,427 hectares) had been converted into forests of multiple-use tree crops with more than 90 per cent ground cover (ref. 6). A variety of saplings of fruit species and fodder seedlings were distributed, and 18 forage centres were established on leasehold forestry land. (The species being introduced in the centres are: Stylo (*Stylosanthes guianensis*), Mulato (*Hybrid brachiaria*), Sterile Setaria (*Setaria*

spp.), Paspalam (*Paspalam attratam*), Forage peanut (*Arachis pintoi*), Elephant grass (*Pennisetum purpureum* cv. Mott and CO3), Ipil (*Leucaena leucocephala*), and Broom grass (*Thysanolaena maxima*).)

Abandoning the traditional practice of open grazing in order to protect the forest was one of the initial challenges that the programme faced. As more and more groups realized the benefits of reforestation and of cultivating a variety of plants, they began to control the grazing of their goats, cattle and buffalo. Many families were instructed on how to build different stalls or pens for their animals, where fodder was brought to them during the day, and the animals were allowed to graze periodically in controlled areas.

**How one group started**

I remember when the forest was open access and I thought that we should not do open grazing, that we should make a plan and form rules and regulations. This is our forest and we should protect it and cultivate it. I had heard about leasehold forestry and also got a book about it. When I read the book I found there were so many provisions and benefits and support from development agencies for income-generating activities. I discussed the idea of leasehold forestry with my community. Not everyone was interested or ready to join. But I tried to convince the rangers to come here to form two groups. The people were divided. They thought that once the government arrived it would seize all of the shifting cultivation plots. But I insisted and went to the DFO to request that two men come.

The hub officer and another officer came to talk to us about the value and benefits of leasehold forestry. The people were then convinced and formed groups. The hub officer provided us with money through the DLO to buy goats. From that money we bought two to three goats. Now, on average each household has four or five goats. Then the programme gave us improved varieties of grass, which now covers the forest. The hub officer visited us regularly and gave us trainings. And 35 of us attended a seven-day training in other cities for forest and forage production.

Now there are more than 15 LFUGs in the area. We have planted banana, pineapple, bamboo, ginger and turmeric, and forestry and forage species. We are poor, but earlier our subsistence was for four or five months. Now we produce enough food to last much longer. And the farmers believe more in groups, and the value and benefits of working together."

*Mr. Lalit Bahadur Praja (Janajati), LFUG chairperson, Raksirang Village (Makwanpur district)*

The latest project-led studies reveal that 44 per cent of LFUGs have strictly controlled grazing and 40 per cent have partially controlled it. In addition, forest fire and encroachment, which are very common in both community forests and government-managed forests, have been totally controlled in about 65 per cent of leasehold forests (ref. 7). These trends are more detailed in the project impact assessment (see Chapter 4).

The LFLP continued similar kinds of plantation work as well as the promotion of NTFPs and distribution of fruit trees. Between 2010 and 2012, more than 10 million seedlings or saplings were planted. Of these, 7 million were broom grass.

"Nature has provided us with the gift of broom grass."

*Mr Jhapendra Bahadur GC, former Chairperson of the District Development Committee, Palpa*

Broom grass was an excellent choice. In Nepal and other countries in the region, especially the vast markets of northern India, there is quite a large market for broom grass (*Thysanolaena maxima*), a special grass used for making brooms for sweeping. The grass thrives best on marginal lands. As well as providing cash income when sold as brooms, it provides green forage for livestock, the roots promote soil conservation, and the dried stems can be used as stakes to support growing vegetables.

**Broom brings boom: how one remote Nepali village has found the road to harmony and prosperity**

In the remote and bucolic Magar settlement of Jhirubas in Palpa district, time stands still. A wrenching seven hour ride from Tansen, the region's Magar culture has been protected till now because of its remoteness: the language, dress, architecture, farming methods, and food.

If Jhirubas was Nepal, the harmony and co-existence here would give us all hope for the country's peaceful future. Magar, Thakuri, and Dalit farmers all trade farm produce among themselves and with their Brahmin and Newar neighbours.

The road has now arrived in Jhirubas and with it has come change. There is great hope that the road will open access to markets for local produce and bring down the price of food and other essentials. There is also potential for tourism: a chance for visitors to trek or bike to an area of Nepal which is still the way Nepal used to be with its quaintly cylindrical red mud houses with thatch roofs.

This region of central Nepal used to depend almost solely on the money sent home by men working in India. But Jhirubas now has another claim to fame: it is a showcase of how leasehold forestry can alleviate poverty.

With help from the District Forest Office [and support from the Leasehold Forestry and Livestock Programme] villagers have converted degraded pastures into plantations of grass used for commercial broom-making. In just three short years, Jhirubas is where the country's largest broom farm is located. The farm has become a model for other villages on how to turn leasehold land into a sustainable income generator.

"It just took the locals here three years to prove that they can achieve anything and they have done the impossible", says Narayan Bhattarai, a government official and hub officer who was instrumental in convincing villagers to lease denuded slopes for broom farming. Being a local who spoke fluent Magar helped and

Bhattarai became a catalyst in convincing villagers to work together for the common good.

Since it was set up three years ago, the 227 households have supplemented their income by producing 3.3 million grass brooms to be sold across Nepal."

Source: Excerpts from *Nepali Times*, 17–23 May 2013, Article written by Naresh Newar during IFAD mission

**Broom grass, stalks and all**

"In the beginning there was disagreement. Why should stop open grazing? This was a problem in the first and second year. But now everyone has realized the value of controlled grazing, once we started earning money from selling our products – especially broom grass.

Thanks to the programme, we have a good surplus, but there is no market. We cannot sell a lot of the surplus. We can only use it for us. We'd like to expand the broom grass business in order to have immediate cash income. We hope to have more support so that we can expand this business. The stalks of the broom grass are not used. We believe that the stalks can be used as raw material for something else.

We would like to know if there are other ways to use this for cash income. One possibility is to turn it into pulp, but to transport it to a factory is too expensive. Maybe we could develop a facility to make semi-processed pulp."

*Mr. Debendra Jung Gurung, LFUG Chairman, Aamdanda village (Tanahu District)*

At the same time, broom grass can cause soil compaction over time, and thus greater water run-off. Therefore, cultivating broom grass is probably best seen as a short- or medium-term income-generating strategy. In the long run, more mixed forest cover is critical.

## The livestock factor

Promoting improved varieties of livestock – mainly buffalo, cows, goats and poultry – was one of the major activities in the livestock component of the first project, the Hills Leasehold Forestry and Forage Development Project (HLFFDP). Altogether 4,852 breeding goats, 414 breeding pigs, and 63 improved breeding buffalo bulls were distributed in addition to 630 buffalo heifers and 412 milking buffalo during the entire project (ref. 12).

The second project, the Leasehold Forestry and Livestock Programme (LFLP), also envisaged increased livestock production as one of the key aspects to reduce poverty of those participating in it. The programme set out to ensure that the livestock component would be initiated only once the leasehold plots and farms had produced enough forage and fodder for goats to be stall-fed.

Goat rearing is a common practice in rural areas, and the main source of income in many households. Therefore, the programme emphasized goat rearing as a means for LFUG members to realize quick benefits. After the lease land is cultivated with forage and fodder, the programme provides a pair of she-goats to each member household and a breeding buck to each group in the second year of group formation.

**Arriving at the livestock component**

This is how LFUGs work for forestry and livestock: identifying groups, group formation, land surveys and training. This is the backbone. The hardest part is identifying the poorest households and forming groups who can plant trees and take care of animals. There are so many households, and the distances are great. We have to go from house to house, to organize meetings. It is very difficult for us to exclude elites from leasehold forestry. It takes a lot of time and negotiations. There are also social political problems. A community can be divided into two or three or four political parties. If there are three different parties and there is the fourth one that doesn't agree with one of the others party, he doesn't want to be in that group.

Once the group is formed, the leasehold areas can be delineated. Some of the areas are productive, while others are less productive or on steeper terrain. These areas are sometimes occupied by customary rule. After much negotiation, the areas are identified and delineated. All of this takes about four to five months.

Looking at the site conditions in and around the piece of land, we ask the people what kind of species they want to plant and what their ultimate goal is, and what product they want to give more priority to (trees, NTFP, grass, livestock). After a series of discussions, the group decides the species and the seasons, after which trainings are provided and a five-year detailed operational plan is drawn up.

The first year is basically dedicated to land development and community organisation, especially to produce grasses so that the livestock component can then be started the following year."

*Mr. Nizam Akhtar, Hub Officer, Munglin*

**A young farmer speaks out**

When the programme arrived, I was very very young, and there was no control over the forest. We simply didn't know. I knew I was poor but I didn't know why. When the leasehold forestry programme staff came and said we would get forest and we could plant, we were astonished to hear this type of thing. We didn't know whether it was good or bad. We thought if the government wants to provide the land and some inputs, let's accept it. We didn't really think it would improve our livelihoods. But we accepted it anyway. When they started giving us trainings and some allowances, we realized that we could really make a difference. We started believing.

At first, very few households were involved in leasehold forestry, but when they saw that people were going for trainings, others soon wanted to join in. For the first five years, progress was slow and the activity of the groups slowed down. But in 2001 or 2002 we decided to stop the open grazing. After some meetings, all of the groups united into one intergroup. I became the chairman of the intergroup. When the community gave me that responsibility, we started creating norms and regulations and even penalties regarding open grazing. People started planting trees and cultivating fodder for livestock. In time, our goat-rearing activities grew. We have started a goat marketing business. We hold a goat fair every Tuesday. People from the city come and buy our goats.

I had just finished my schooling when the programme was starting to show results. I was a young chap. I stayed here. I can do something for my community. I must continue the work here and help motivate others."

*Mr. Som Bahadar Magar, young member of an LFUG, Shaktikhor, Chitwan District, site of the first leasehold forestry initiative*

As of July 2014, a total of 79,134 breeding goats (plus an addition 1,189 from the Technical Assistance) and 4,158 breeding bucks (plus an additional 139 from the Technical Assistance) had been distributed. Through this initiative, the number of goats has increased in 75.8 per cent of the LFUGs, remained the same in 18.4 per cent, and decreased in 5.7 per cent.

**Table 5: Goat rearing status**

| *Indicators* | *# group* | *%* |
|---|---|---|
| Increased | 3,824 | 75.8 |
| Remain same | 929 | 18.4 |
| Decreased | 289 | 5.7 |
| *Total* | *5,042* | *100.0* |

The programme also promoted the exchange of breeding bucks, which should be done on a regular basis to avoid in-breeding and improve the quality of the goats. Breeding bucks should be exchanged every 12 to 18 months. This initiative has shown mixed results: breeding bucks are exchanged within 12 to 18 months in 30.2 per cent of the LFUGs, within 18 to 24 months in 28.3 per cent, and not exchanged in 41.5 per cent.

In Palpa District, for example, a Goat Resource Centre was established involving 80 farmers, who were coached in improving 80 goat sheds. Ten bucks were added to ensure reproductive effectiveness in the resource centre, and a "performance rating system" was set up for the breeding does. In addition, one group member selected by the intergroup was trained as a Village Animal Health Worker, and he went on to establish a veterinary medical shop. A goat dipping tank was also constructed to rid goats of external parasites (ref. 6).

**Keeping livestock healthy**

"I was nominated to attend a one-month livestock training. During the training I learned all about vaccinations and primary symptoms of different diseases, medicines and a special training for when buffalo are giving birth. People were reluctant to use vaccines or to follow the instructions. It was a very hard time for me, but slowly I convinced the farmers to use them, and to use them properly. Now they understand how beneficial the vaccines can be. The district offices used to provide the services for a very nominal fee. But for the past four years this type of service has not been offered."

*Young man (unnamed), member of an LFUG in Darechowk village (Mungling/Chitwan District)*

Animals should be vaccinated every year against diseases and treated every six months against parasites and *peste des petit ruminants* (PPR) to keep them healthy. In this regard, the programme has trained and created 146 Village Livestock Assistants and 161 Village Animal Health Workers. Some of the latter also received loans ranging from NPRs 15,000 to 18,000 (USD 150–180) as a revolving fund to start up veterinary services in their areas. In addition, the Village Animal Health Workers and 1,121 LFUGs participated in seven-day training on goat management and health. And Junior Technicians working in the field received training in grass and seed production, animal feeds, breeding and social mobilization.

**Rising in the ranks, how I became a veterinarian**

"I was not born in this area. I came here from another ward many years ago. I happened to be here during the formation of the first leasehold forestry users group. I remember that most of the people were frightened of the forestry officials and thought they had come to arrest them or take their land. They would come to the first meetings and then run away, but I felt I had nothing to lose and I would stay.

One day, one of the forestry officials came to me and asked me who I was and if I was poor or not. I told him my family was very poor. We were eight brothers and our father only owned half a hectare of land. This was the reason why I had left my home. He said I could join the leasehold forestry programme if I wanted. He also told me he thought I was very clever and creative and asked if I wanted to help form one of the groups. I decided to participate and was soon elected chairperson of my group. I am still the chairperson today. I then took a 45-day training away from the village to become a village animal health worker. I was so interested and committed that when the course ended the line agency service providers lent me 4,500 Rupees (USD 45) to help me start my own business. I had also saved 4,000 Rupees (USD 40) from the training allowances. The veterinary director of the course was pleased with me and asked the wholesaler to give me 10,000 Rupees (USD 100) worth of medicines. The director gave his guarantee that I would pay the amount back soon. That was when I started working as a veterinarian and supplier. My main activity was castrating goats and bulls for the members of my LFUG. But in time I met many people from many different places and rapidly expanded my business. I decided to take trainings in agronomy and when I was competent enough I took a loan from the Agricultural Development Bank and opened a shop where I sell agro-seeds, fertilizers and pesticides.

I'm completely independent from the programme now, but I always give priority to leasehold forestry users' group members and make sure they can afford my products. I think many LFUG members know me and trust me now, and they come to my shop from many areas. I don't have the time to visit all my customers, but when they come I try to give them the best advice and the latest and most effective medicines. The shop is not the only social service I provide. Because of the training and the experience that followed, I'm the LFUG intergroup chairman, the Secretary of the Health and Sanitation Committee, a member of Nepal's Red Cross Society, a member of the Chamber of Commerce and the Treasurer of the Credit and Saving Collective. I have learned a lot and come a long way. But all of the members of my group are now richer and happier because of the leasehold forestry programme, and I believe our group is very strong. I'm so thankful to this programme because it helped me become what I am now. I am also thankful to this community because I've been accepted as if I was born here."

*Hari Sheran, shopkeeper, veterinarian and chairperson of one of the leasehold forestry user groups in Raksirang, Hetauda District*

Vaccination and drenching are in regular practice in 22.2 per cent of LFUGs. Either vaccination or drenching is practised regularly in 37.9 per cent of LFUGs. No vaccination and drenching are in practice in 39.9 per cent of LFUGs.

**Table 6: Animal health services**

| *Indicator* | *# group* | *%* |
|---|---|---|
| Vaccination and drenching regular | 1,119 | 22.2 |
| Either vaccination or drenching regular | 1,909 | 37.9 |
| No vaccination and drenching | 2,014 | 39.9 |
| *Total* | *5,042* | *100* |

Goats continue to be the prime source of income (24 per cent) for most of the households. Moreover, a recent FAO study on goat marketing in three of the programme's districts reveals that on average the annual income from goats varies between NPRs 46,000 and 145,000 per household (USD 460 to 1,450), with a net average benefit of NPRs 56,633 per household (USD 566) (ref. 9). Similarly, a site-specific study in the Mugling leasehold forestry cluster in Chitwan District shows that number of goats increased by about 43 per cent during the 12 years of the leasehold forestry intervention, and the annual income of LFUGs increased from NPRs 6,000 to NPRs 18,000 per LFUG (USD 60 to 180) (ref. 27).

**Promoting improved animal husbandry practices – some facts and figures**

- 22 people have been trained as Village Animal Health Workers to supports LFUG members.
- 19 Village Animal Health Workers have been trained to run their private agro-vets with business plans and to effectively deliver livestock health services to LFUG members.
- 967 users (57 per cent women) have been trained on goat rearing and shed improvement through 48 training events.
- 3 model goat sheds have been constructed as demonstration sites in Chitwan, Gorkha and Lamjung districts.
- A study entitled "Economic goat herd size for LFUG members" has been conducted.
- 11 "Animal health campaigns" have been organized in coordination with the District Livestock Service Office. More than 4,500 goats and cattle have been treated during the campaigns.
- A study entitled "Goat Insurance Scheme" explored financial instruments for safeguarding household assets in case of unexpected animal mortality.

Source: FAO/IFAD, December 2012. Technical Assistance for Leasehold Forestry and Livestock Programme. Key progress 2010–2012 (GCP/NEP/062/FIN)

Some households have started commercial goat-keeping and others have shifted from goats to larger animals (milking buffalo or cows). This is a natural and positive evolution, and reflects the dynamics of rural livelihood development.

## Money matters – rural finance

Rural finance is the third element, after forestry and livestock, of the components of the LFLP. It was set up in order to promote savings and credit schemes among LFUGs and help them create rural financial institutions known as leasehold forestry cooperatives (originally called Village Finance Associations).

The experience of the HLFFDP in its involvement with the Agricultural Development Bank had not been very promising (the payback rate was only about 48 per cent) because of two reasons: The Maoist rebellion had started and they forced the users not to pay back; and a "grant syndrome" took hold among the borrowers. They believed that since they were poor and the money was provided to them to alleviate their poverty, it was incumbent upon the government to shoulder the burden of the loan. Given this experience, the LFLP designed a member-based rural finance mechanism, but at the same time realized that it needed to build group capacities to promote efficient and sustainable use of funds, and to establish an engrained mechanism of self-sponsored member-based "Saving and credit".

To enhance the capacity of the groups, social mobilizers were used, to whom the TA had provided training for this purpose. The cooperatives were registered as Savings and Credit Cooperatives. An innovative internal financing mechanism for cooperatives was developed, providing term deposits from the programme's goat grants. One eligibility criterion for these grants, provided instead of in-kind loans, was the pledge by LFUG members to deposit NPRs 1000 (USD 10) into the cooperative's fund within 24 months of receiving a goat, and to keep the deposit in the fund for three years. The money could be used as initial seed money, but the provision was that they were free to take the money back with the earned interest if they liked, or keep it in the cooperative as their own saved money. In the cooperative statutes, members agreed to make these deposits available for on-lending, provided that institution met certain criteria to ensure it conformed with minimum levels of sound performance, including a portfolio at risk (>30 days) ratio of less than 5 per cent. Such a financing mechanism enabled cooperatives to operate entirely on internally mobilized funds, at least for their initial years, and contributed to a strong sense of ownership and a high-quality portfolio.

There was also a need to develop institutional arrangements that were stable, could build assets, protect against risk, and support all the potential income-generating activities. Different methods were explored, such as group lending and liability, pre-loan savings requirements, loan sizes and guarantees that comprised microfinance institutions. Considerable technical support and knowledge from other parts of the country was needed to initiate such arrangements, and to refine them to meet the requirements of individual groups. A local NGO (Fri-Pad) provided technical support to form 36 new cooperatives or Village Finance Associations. These cooperatives were registered with the appropriate authorities as formal institutions under the Cooperative Act 2048. Fri-PAD then provided technical support to LFUGs on topics including mass orientations, cooperative education, mass meetings, policy and bylaw preparation and other documentation. Fri-PAD also guided LFUG members in becoming shareholders in eight other cooperatives.

Another NGO, COCIS, formed additional cooperatives and provided training geared toward different types of cooperative members. Training topics included: cooperative management for executive members; book-keeping and financial management; micro-finance (loan management); business plan preparation; NTFP-based micro-enterprise development; observation study tours; and market linkage development workshops.

**Local institutions and microfinance as the key to sustainability**

"In order to sustain our efforts we are developing a third phase, which will have microenterprise development as a main component. The first tier of this component is through group savings and credit schemes. The second tier is through cooperatives. This is the way to sustainability. Groups need to form cooperatives, with 100–200 people. We hold fast to this cooperative concept. After three years of their own management, they are supposed to link with external microfinance services. We have many of these services here. Banks need to inject their money into microfinance. This is mandatory by the government. Cooperatives could be linked in this way if they are mature enough and have better management. The third phase is in a planning phase until September 2014. We have hired a consultant team to prepare a kind of project document for this phase. It will include these second-generation issues of LFUGs as well as climate change issues. This will be taking place over the next few years, and much of it will be done in the same programme areas. Livelihood improvement was the first phase. Now we can move up. We need to go forward, to move ahead. And that is where microfinance and diversification of income come in to play.

If we are not there, then what? We think these cooperatives will be the institutions that will replace our work and our activities. But this is lagging behind. In total there are 56 cooperatives of LFUGs. We need to prepare some 3,300. Local institutions are the only option. They need time."

*Govinda Prasad Kafley, Team Leader, TA-LFLP*

Eventually the Technical Assistance programme took over this support. Local resource people were hired in three clusters (Dandeldhura region, Dolakha region and Munglin region) to support the cooperatives. The Technical Assistance programme also conducted an analysis of the status of cooperatives, which revealed that the capacity of the cooperative had been enhanced: two cooperatives were able to access funds from the central bank, and eight were in the pipeline. The others remained self-help groups at the LFUGs level.

Saving and credit schemes are almost institutionalized in leasehold forestry. More than 90 per cent of the LFUGs have set them up, and more than 75 per cent of the groups are actively involved in them. All the members of a group deposit a fixed amount as monthly saving. (The amount varies from one group to another.) The average monthly saving rate of LFUG members is NPRs 12.6 and ranges between NPRs 5 and NPRs 200. Of the total LFUGs participating in the schemes, about 4 per cent have total savings of more than NPRs 50,000 (USD 500), about 37 per cent have total savings between NPRs 10,000 and NPRs 50,000; 59.5 per cent have total saving of less than NPRs 10,000.

LFUGs mobilize saving amounts as loans to their members. The amount of group saving and purpose of the loan determines the loan amount. The loan amount varies from one group to another. LFUG members take loans to meet their emergency needs and undertake income-generating activities. A total of 33.5 per cent LFUGs have used over 90 per cent of their savings for loans, 13 per cent have used 75 to 90 per cent of their savings for loans, and about 54 per cent have used less than 75 per cent of their savings on loans for various income-generating activities (Ref. 8).

> "We earn money from broom grass, and each household pays 20 rupees per month into our savings and credit scheme. With our savings, we bought a tractor. There are 7,000 LFUGs in Nepal, and ours is the first group to buy a tractor!"
>
> *Unnamed, LFUG member, Khader, Jhirubas*

As mentioned before, each household that has received goats from the programme is supposed to deposit NPRs 1000 (USD 10) per goat in the group account within 18 to 24 months as a term deposit – a mechanism developed to compensate for the absence of seed money. The study shows that in 27.7 per cent of the LFUGs, all members of the group have made the deposit; in 5.8 per cent of the LFUGs, only a few members of the group have done so; and in 66.6 per cent, none of the group members have made the deposit.

Not all was smooth sailing. If all the money deposited by the individual groups was to be collected in the cooperatives, the lower tier of saving and credit would be functionless. One of the main reasons for the group's monthly meetings was for savings and credit matters, which provided loans easily and safely. The groups did not want to transfer their money to the cooperatives, thus having it rest in the hands of others. In addition, as the cooperatives were newly formed, they did not have experienced personnel to run them, or sufficient funds to spend on administrative matters such as salary and supplies. Therefore, they were run by a skeleton staff who was not experienced. Ten groups with an average of nine households per group made 90 members for the cooperative, a critical mass that was not sufficient for the cooperative to meet economies of scale. In those instances when the operation was larger, with more users and greater amounts of money, the transparency and integrity of the system could be compromised – with elite capture occurring. In fact, at one point, a review mission was so critical that it provided recommendations for dismantling the cooperatives and distributing the money to the respective owners. Fortunately, many of the weaknesses were addressed, and most of the cooperatives had smooth operations.

**Figure 5: Status of goat term deposit**

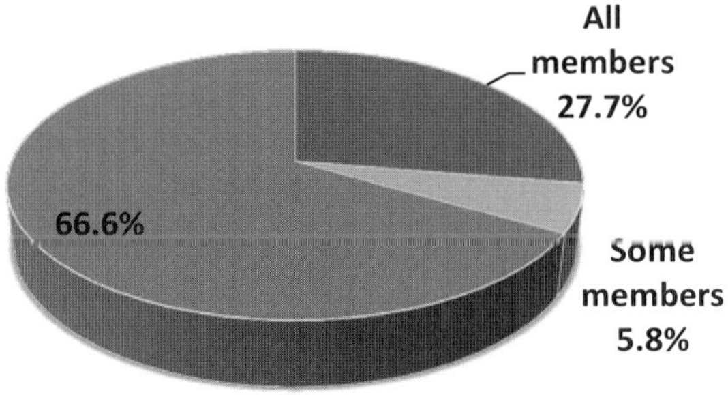


Source: ref. 8

## Advancing to a higher level: cooperatives and federations

**A Cooperative in Tanahu District**

"Our cooperative started about 13 years ago. There were 20 LFUGs and 35 shareholders originally. Now there are 153, and 103 of them are women. We started by collecting milk, but now our cooperative is about mobilizing funds and running a savings and credit scheme. Cooperative members pay monthly dues of 100 rupees (USD 1). They can deposit more if they like, and if they do they earn extra interest. Initially the dues were 20 rupees per month, then 50 and now 100. We are thinking about raising them to 500 rupees per month (USD 5). But the decision must be a unanimous one and based on the situation of each member.

If one of us wants to have a loan, at least seven members need to provide a guarantee. Then the cooperative leaders evaluate the person, the amount of the loan and the reason for it, and whether it can be repaid. Based on this, a decision is made about providing the loan. The maximum loan period is one year, although sometimes extensions are given since not everyone can pay back the loan in that period. Our repayment rate is more than 60 per cent.

I myself took a loan to purchase a cow, which enables me to sell milk. Things were going so well that I took a second loan for another cow. Others have taken loans to buy goats, construct houses, send their children to school, open grocery shops, start vegetable farming, and so on. Only three members have not taken loans.

We want to see our cooperative get involved in different activities – commercial farming and other enterprises. We want products to come here so that we can sell them through the cooperative. Each member needs to be creative and innovative. If each household could produce one product and bring it here to sell, we would go a long way. The support we need is the knowledge and technologies for commercialization, as well as coordination on market linkages. And then it is we who can create our own development, especially local employment so that the young people will not migrate. We poor people are united in that we can improve our livelihoods. That is our vision. Yes, there are a lot of challenges. But we hope that donors, local partners and the government will join hands with us."

*Chairperson of the Cooperative, Tanahu District*

The trend of networking savings and credit groups into cooperatives to meet the services of rural finance institutions is also steadily increasing. The number of cooperative increased from 19 in 2005 to 54 in 2012. These 54 cooperatives represent over 5,000 savings and credit group members from almost 500 LFUGs in 15 districts. The total average saving balance of these cooperatives is NPRs 2,131.0 per shareholder (NPRs 2,202 for women, USD 22), and the average outstanding loan balance is NPRs 10,243 per shareholder (USD 102). The recovery rate is 93 per cent. In addition, cooperative earnings were about NPRs 1.2 million (USD 12,000).

However, the overall portfolio quality of these cooperatives is still below the global standard and practices. According to the Cooperative Assessment carried out by the FAO, out of the 54 cooperatives reviewed only 19 had demonstrated some limited potential to develop into viable institutions. The use of member savings by cooperative leaders is one of the most important issues.

The approach proposed by several supervision missions is to link LFUGs to existing micro-finance institutions and to the Poverty Alleviation Fund. About 30 per cent of LFUGs have established links with several different financial institutions. Nevertheless, with the right kind of technical support for their management, growth and development, most of the cooperatives can become viable and sustainable institutions (ref. 8). The case studies that follow illustrate this potential.

**Cooperatives and LFUGs in Kavrepalanchok District – a typical story**

Rayale is a village of nine wards with around 4,000 households in Kavrepalanchok District around 20 km southeast of Kathmandu. This is the region of the "middle hills" of the Mahabharat Range, also known as the Lesser Himalaya. The Middle Hills are densely populated in the lower valleys while population thins out above 2,000 metres. Paddy- and cereal-based agriculture is very common in the lower valleys, whereas cold-tolerant crops such as potatoes are cultivated in the higher valleys. Rayale has paddy fields along the river valley floor, non-irrigated cereal fields higher up, and forests on the highest slopes, on which much of the income-generating activities are based.

The Leasehold Forestry and Livestock Programme started supporting the formation of LFUGs in the district in the early 1990s. Seven LFUGs were created, and the cooperative of Kalpabriksha resulted from the consolidation of some of those groups in 2011. The villagers have begun a number of income-generating activities, including milk and dairy production, and vegetable cultivation (cooperatives) and cardamom, harvesting of grass and agro-forestry management (LFUGs). Women are playing a key role in developing new income-generating activities within their households and at the cooperative.

The cooperative has 214 members, the majority of whom are women. The two main functions of the cooperative are to provide loans and technical assistance to its members. To become a member NPRs 1,205 (USD 12) is required to cover entry fees and administration costs. The capital of the cooperative reaches NPRs 126,000 (USD 1,260), and members can take loans at an interest rate of 14 per cent, which increases 2 per cent each month that the loan is not paid back. Meetings are held on the 10th of each month, and a general assembly is organized once per year. "Transparency is the most important part of the cooperative", the cooperative's treasurer said.

In addition to the activities that the cooperative already manages, the members are very interested in developing tourism in their village.

The president of one of the LFUGs, who is a member of the cooperative, received training by the DFO, and then received maize seeds to plant. He also took a NPRs 5,000 (USD 50) loan from the microfinance institution to purchase goats and buffaloes. The interest rate was 16 per cent. If a borrower pays the loan back on time, the interest rate decreases for any subsequent loans.

Although he was able to improve his milk production with the new animals and owing to the fodder he was able to cultivate, it was not profitable enough (NPR 70,000 or USD 700 per year). He started to make silage in order to feed the animals during the dry season, an innovative technique for the area.

Another cooperative member has a dairy milk farm, which his family started 14 years ago. The idea was to establish a place to collect milk from neighbouring farms. Any farmer living within a one-hour walk can deliver the milk to this dairy.

At the beginning production was very low, with only 30 litres collected each day. Today the farm has the capacity to collect 400 litres each day. The quality of the milk has also improved since they acquired an improved breed of cow that produces milk with a higher fat content. The price of milk has also increased a lot.

"Milk production is not a priority for the government of Nepal", said the father who built the dairy.

The dairy is not isolated. About 10 million litres per day are produced in the district and 7,000 litres per day in Rayale. However, the dairy does not sell milk at the national level because of stringent and costly requirements. Instead, they are constrained to sell their milk to the private sector without any real bargaining power.

The farm has to cope with other challenges as well. The main challenge is the high death rate among the cows. Another is the decrease in the amount of milk produced during the dry season.

Another cooperative member produces cardamom, an initiative he took on his own. The first year, the programme provided him with 200 plants. Today he has 52,000 plants in his field and sells the fruits and seeds in the market.

He said that patience is the most important thing, explaining that the first three years were unproductive. He added that production of cardamom is for those who can invest and wait some years before realizing any profit. But then this is a very profitable plant, and does not require a lot of care. Only some irrigation is necessary.

**Cooperatives in Kavrepalanchok District – another typical story**

It is early in the morning in Kavrepalanchok, Rayale in the hilly District of Kavre, Central Nepal, about a five-hour drive from Kathmandu. Lekhnath Sapkota is busy preparing his daily work, which is quite full. Running a local dairy cooperative called Phulchoki, which is named after a temple in his area, he is buying milk from local farmers that is then put in a chilling system to be preserved. The milk will then be purchased by a commercial buyer who goes directly to his community to help transport it to dairy factories for processing.

The cooperative started off with a daily collection of only 30 litres but is now able to buy up to 500 litres of milk each day. With little land for cultivation, this provides regular work opportunities for around 80 households in the community that raise water buffalos or cows.

"I am very happy because as a farmer I can support other farmers with income generation activities that they can do on their own and based on their potential. On average, we buy 8–20 litres of milk from each household every day", said Sapkota with a smile.

In a small village like Kavrepalanchok, an extra of a few thousand rupees goes a long way, giving local farmers extra cash to buy school textbooks for their children or to re-invest in cattle-raising.

"In the past, we sold milk individually to buyers and they decided the price. Now we can be in a much better position to negotiate the price and maximize the profit from selling in bulk quantity as a group", continued Sapkota.

Bhim Bhadur Timilsina is a local farmer who has been supplying milk to Phulckoki for almost 15 years. The 50-year-old farmer noted that his household is very content with the reliable source of income from selling milk.

"I receive money from selling milk twice per month of 15,000 to 17,000 rupees (USD 150 to 170) and payment is never late", he said. "I am sure that I and other farmers will continue selling milk to Phulchoki in the future. Together, we will achieve more as a group."

He is now looking at loan possibilities from the Government or international organizations in order to expand his business. Dreaming of building his own dairy factory, he hopes to go into large packaging and commercial production of dairy products for local markets, offering many more jobs to local farmers.

Source: Asia.ifad.org Knowledge Management community April 8, 2013. Written by Cecina, IFAD Vietnam Knowledge Management and Communication Specialist, participating in PROCASUR learning route.

**An assessment of cooperatives: much still needs to be done**

In 2012, the Technical Assistance project for Leasehold Forestry and Livestock Programme commissioned CED-Neal, a national NGO, to conduct an assessment of 54 cooperatives receiving financial support from the programme. Out of the 54 cooperatives, 2 are good, 16 are fair, 28 are poor, 6 are very poor and 2 are informal. Some of key findings resulting from the assessment are as follows:

- The cooperatives are mobilizing a significant amount of local savings. However, members' savings are very small (NPRs 2,131 per household or USD 21).
- Access to loans by shareholders/members is less than 30 per cent, and about 10 per cent of loan clients default.
- The average outstanding loan balance is NPRs 10,243 (USD 102).
- The overall portfolio quality is below the global standard and practices: recovery rate is 93 per cent; portfolio in arrears is 6.3 per cent; and portfolio at risk is 24.7 per cent.
- The cooperatives extend loans to Board of Director members; in more than 50 per cent of these cooperatives, the loans are not being repaid. In some cooperatives, outstanding loans to Board of Director members exceed 20 per cent of the total loan portfolio, which puts the savings of the shareholders at risk.
- Management of assets and liabilities is quite new among these cooperatives. The Boards of Directors and staff of most cooperatives are not confident about asset and liability management. Moreover, the loan management system is not particularly scientific. Although there are loan committees to manage the portfolio, the Boards of Directors have a dominant role in approving and disbursing loans. Thus there is little if any effective control by cooperative members over what happens to their savings.

LFLP should provide selective support to leasehold cooperatives: extensive technical support to good and fair cooperatives; and intensive support to poor cooperatives to ensure their sustained growth and development. LFLP should consider dropping very poor and informal cooperatives from its support package.

Support to the 18 cooperatives considered acceptable should include: building effective management information systems, promoting savings and diversifying savings products, offering services that fit the needs of the clients members, simplifying procedures to reduce operational costs, motivating clients to repay loans and focus on high repayment, charging adequate (above market) interest rates and fees to be sustainable, involving clients in designing services, promoting effective governance, and focusing on financial services only and developing institutional linkages.

Source: ref. 4

These difficulties for self-help groups – especially when they become larger and more complex like cooperatives – to properly manage savings and credit are symptomatic of the rural finance market in Nepal. On one hand, the traditional banking systems remain in large cities, leaving the ground to traditional money lenders in rural areas and their extraordinary lending rates. On the other hand the emerging micro-finance system is slowly spreading throughout the country. From this point of view, self-help groups involved in microfinance are an excellent way of getting people sensitized to the use of savings and credit. But this becomes very limited as soon as the loans are used for professional activities and the development of private business, as the self-help groups do not have the banking background needed. Self-help groups should be encouraged to join microfinance cooperatives, which in turn can join larger banking systems, but they need considerable support to grow organizationally and minimize their financial risks.

## Overall performance of LFUGs

The various IFAD annual supervisions and the FAO study conducted in 2012 included an overall assessment of LFUGs, based on four aspects of performance: institutional development; leasehold forest development; livestock development; and rural finance. The study showed that 21.9 per cent of LFUGs were considered active, 57.2 per cent were considered moderately active and 21 per cent were considered passive. Even among groups that had been formed during the first project and operating for 15 years, still 62 per cent were considered active or moderately active.

**Figure 6: Overall status of leasehold forest user groups (in %)**

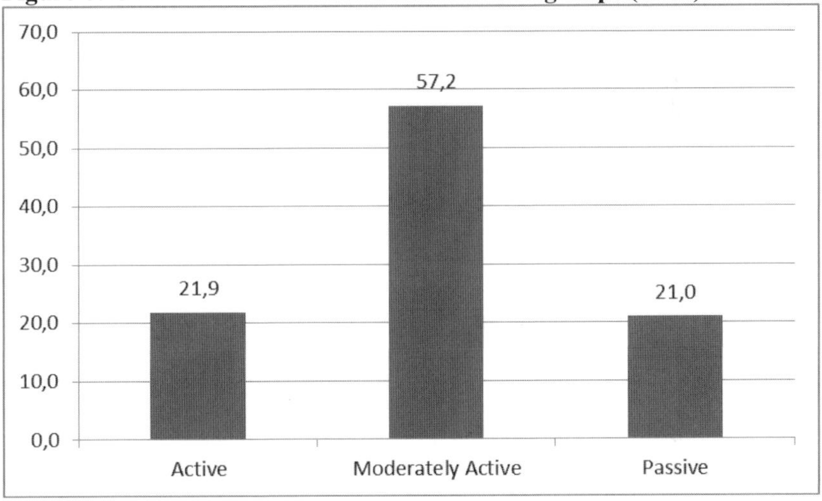


Among group activities, leasehold forestry activities scored highest, with acceptable performance in 89 per cent of the groups. Field visits confirmed this information: group members were categorical about the benefit of having obtained an asset for a period of at least 40 years. These indications point to the importance of the leasehold concept in that it provides a long-term productive asset to poor households, who actively maintain the institutions (groups) related to it (ref. 20).

**Figure 7: Performance by component**

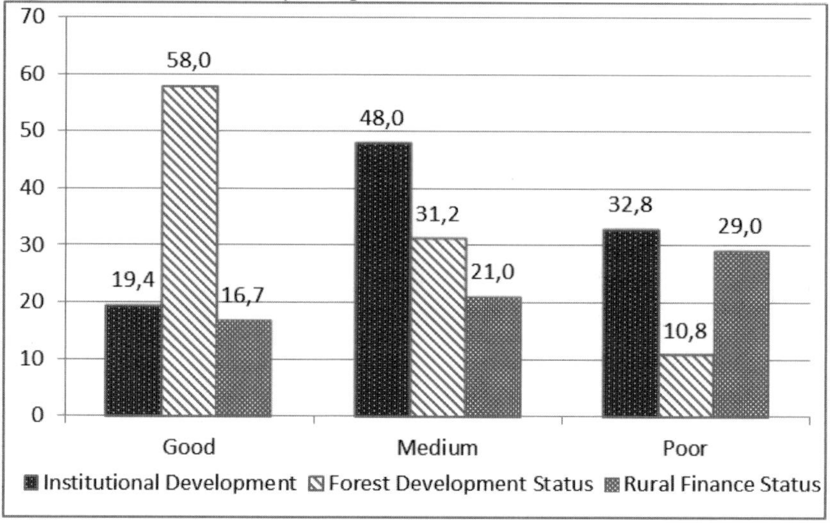


When the overall performance of LFUGs in the first and second project is compared, the first phase shows a negative trend: The percentage of active or very good LFUGs decreased about 5 per cent, from 23 per cent in 2005 to 17 per cent in 2012. Similarly, the percentage of passive groups increased about 14 per cent, from 25 per cent in 2005 to 38 per cent in 2012. The rankings were based on: institutional capacity; forest and biodiversity; livelihood improvement; and gender and equity.

Similarly, an analysis of 3,374 LFUGs from the phase two project (including the groups formed during the bridging phase) shows that 24 per cent were active, 63 per cent were moderately active and 13 per cent were passive.

It is important to note that a considerable percentage of LFUGs from the phase one project are still active after more than ten years, which is quite an achievement. Historically, when a project closes, the tendency is for groups formed with the support of the project to collapse soon after – and this has been far from the case with the project.

**Table 7: Leasehold forest user group performance: comparison between first and second project**

| | *First project* | | *Second project* | *Overall* |
|---|---|---|---|---|
| *Group performance* | *2005/06* | *2011/12* | *2011/12* | *2011/12* |
| Active | 22.7 | 16.9 | 24.3 | 21.8 |
| Medium | 52.7 | 45 | 63.2 | 57.2 |
| Passive | 24.6 | 38.1 | 1.4 | 21 |

Source: After DVN/NPC, 2005, TA FAO, 2012

**Summing it up in numbers**

- About 38 per cent of LFUG members are women. The representation of *Janajatis* in LFUGs is highest (58.8 per cent) among the ethnic groups. About 21 per cent of LFUG members are *Janajati* women. *Dalit* represents only 12.5 per cent of the total LFUG members. Out of this 6 per cent are *Dalit* women.
- Group meetings are regularly conducted in 63.4 per cent of LFUGs. In 22.4 per cent LFUGs, group meetings had not been organized in the last six months.
- The role of women in household decision-making has increased substantially. Decisions by women alone rose from 10 per cent to 25 per cent, and decisions taken by men and women together rose from 30 per cent to 55 per cent. Consequently, decisions by men alone declined from 60 per cent to 20 per cent.
- About 72 per cent of leasehold forests are well managed and 5.2 per cent are poorly managed.
- In some sites, vegetative ground cover has increased from 32 per cent in new sites to over 90 per cent in about seven years.
- Grazing is totally controlled in 44 per cent of leasehold forests but it is still open in 16 per cent.
- Forest fire and encroachment are totally controlled in 64.8 per cent of leasehold forests. Both are still very common in 8.9 per cent leasehold forests.

- Forage and fodder are available for more than 6 months in 13.4 per cent of leasehold forests but are available for less than 3 months in 41.6 per cent.
- About 34 per cent of LFUGs have mobilized over 90 per cent of their savings, whereas 53.5 per cent of LFUGs have mobilized less than 75 per cent of their savings.
- The number of goats has increased in 75.8 per cent of LFUGs, remained same in 18.4 per cent, and decreased in 5.7 per cent.
- The breeding bucks are exchanged within 12 to 18 months in 30.2 per cent of LFUGs. They are not exchanged in 41.5 per cent.
- About 58 per cent of leasehold forests are well managed, and 10.8 per cent are poorly managed.
- The livestock development status is good in 43 per cent of LFUGs. It is poor in 21.8 per cent.

Source: Ref. 7, ref. 20

## Open issues

"The leasehold forestry programme has a number of significant challenges. The land that is handed over is degraded forest, and a lot of funding is needed to improve that land. In developing the land use plans, an extraordinary amount of time and effort is required to work with poor people, and each plan is very context-specific. I would say that the time element is even more critical than the funding element. For these people, we have to give so much time. Even more than money. And there is always the risk of creating a dependency syndrome, which is greater the further west you go."

*District Livestock Officer, Tanahu*

The leasehold forestry experience has undoubtedly had its share of success. Huge swaths of land have been transformed from barren slopes to fertile and productive forest; thousands of poor marginalized rural people have risen out of poverty, and have taken on the skills to ensure that they don't slip back. It is safe to say that leasehold forestry is here to stay. The question is how to capitalize on the experience, making sure that the initiative is even better as it moves forward. In this regard, a number of issues emerged along the way, which should be addressed if leasehold forestry is to have an even greater impact in the future.

**Equity**. Some poor households are not aware of the leasehold forestry programme, and not all of them receive prior information about the selection of LFUG members. In many cases, this is reinforced by their remote geographical location and poor infrastructure, which limit the flow of information and mobility of the people. As a result, some middle class and even wealthier households are included in the LFUGs, which was not originally intended. At the same time, there is a trade-off between dogmatic equity considerations and socially feasible targeting approaches. At the same time, "elite capture" has become only sporadic, occurring in only a few cooperatives.

**District Forest Office implementation capacity**. Leasehold forestry is primarily implemented by the Department of Forests (DoF), which also implements other community-based forestry management programmes. For the District Forest Offices (DFOs) and their forest rangers, the implementation of another programme means additional work. With no additional permanent staff added to implement leasehold forestry, the DFO and staff have the challenge of juggling their time and resources to accommodate the programme. This often results in shortcuts in the processes, which consequently lead to lower-quality outputs. Given the nature of their incentives, the line agencies will invest more of their time and resources in the programme when there is greater funding. In other words, they may invest more time initially, but after funding wanes, so does their support. Indeed, very much is demanded of DoF staff at the district level, and there is a wide disparity in employment conditions between government staff and staff hired by grant-funded programmes and NGOs. Daily allowances are less than a quarter of what is paid by grant-funded projects. Study tours and in-service degree courses are much appreciated but opportunities are rare. Range posts lack amenities and equipment, and rangers are required to be responsible for excessively large territories. This is an incentive for the government to increase budget support to leasehold forestry in the districts.

**Prohibition of the cultivation of vegetables and cereals**. Under leasehold forestry, the cultivation of vegetable and cereal crops is legally prohibited; only grasses, fodder, and trees are allowed. For the poorest who live a hand-to-mouth existence, growing grasses or trees may not seem rational because it would take months or even years before they are harvested. On the other hand if they are landless, once the land being cultivated by these poor farmers is identified as degraded land, they would have not much choice but to hand it over since they do not have ownership rights to these lands.

**Limited bargaining power of LFUGs**. The current leasehold groups range from 5 to 15 households, which is very small compared to community forestry or other farmers' groups. This is a disadvantage when accessing external support since NGOs and other civil society organizations often look for larger community groups with which to collaborate. There is also the problem of registering these smaller groups with the District Agriculture Development Office and other district line agencies, which require larger membership. The creation of federations of LFUGs is a potential solution.

**Monitoring and evaluation**. In principle, the leasehold forestry process of handing over forests to the poorest tries to minimize exclusion – it is for this reason that the programme was designed and implemented in the first place. However, as mentioned, exclusion does occur, and this can be attributed in part to how programme implementation is monitored and outputs evaluated.

Although participatory monitoring and evaluation is supposed to be one of the features of the programme to show its impact, this is in its early stages of development and roll out.

Moreover, programme outputs are mostly measured based on quantity (e.g. number of households organized; area of degraded forests covered) rather than on quality (e.g. number of households rising above the poverty line; sustainability of the livelihood projects implemented and eventually of the LFUGs). Because of the emphasis on quantity outputs, the line agencies often do not follow the ideal but lengthy process of implementation, which takes into consideration the uniqueness of certain groups, such as their ecological, geographic and socio-economic conditions.

**The inherent nature of poverty**. Many of the issues mentioned can be attributed to the fact that the people being targeted and organized are the poorest. Being poor, they lack various resources or assets to participate in activities other than those that will give them immediate returns. In addition, they also lack the power to prevent other people from obstructing them from claiming their rights or privileges such as the ones being provided through the leasehold forestry programme. They are often illiterate, live in remote areas, and are indebted to the better-off households in their neighbourhood. Therefore, the sustainability of LFUGs could be an issue, although, as previously mentioned, the Group Assessment study indicated that groups remain active long after the closing of the project. These groups (particularly the genuinely poorest) need more than just two years of institutional and technical support as well as monitoring. In this regard, the role of Group Promoters and the DFO and DLO are critical (ref. 1).

## A word about "Protected Areas"

The management regime of national forests in some districts (e.g. Chitwan, Makwanpur, Sindhuplachowk) where leasehold forestry was being implemented has been changed with the declaration in 2004 of buffer zones and conservation areas. Now the jurisdiction of leasehold forestry falling into the territories of these protected areas lies with the concerned Warden of National Parks and the authority of conservation areas. A total of 330 hectares of leasehold forests previously handed over to 85 LFUGs under the Forest Act 1993 and Forest Regulation 1995 were included in the buffer zones of Chitwan National Park, Parsa Wildlife Reserve, and Gauri Shankar Conservation Area with the declaration of buffer zones and conservation areas. Neither the National Park and Wildlife Conservation Act 1973 nor the Buffer Zone Management Regulation 1996 and Conservation Area Management Regulations 1996 have provisions for pro-poor leasehold forestry inside buffer zones and conservation areas. However, LFUGs inside buffer zones persisted, managing the leasehold forests until the end of the operational plan period, based on the provisions of the Buffer Zone Management Regulation 1996. They were not permitted to harvest or use the forest products of the leasehold forests once the operational plan period was over.

To facilitate the implementation of leasehold forestry inside buffer zone and conservation areas, a draft provision for leasehold forestry in buffer zone management regulations was prepared by TA-LFLP and submitted to the Department of National Parks and Wildlife Conservation.

The high emphasis on conservation and poor service delivery mechanisms have seriously distressed some of the LFUGs whose land is part of the protected areas system. They have been prevented from implementing their operational plans, from harvesting forest products, and most importantly they have become functionless because of severe delays in the revision/renewal of their operational plans. At the same time, their motivation has plummeted, as they have no idea whether the new tenurial arrangement will provide them the same rights in managing and utilizing the forest products as they had enjoyed earlier.

**Protected Areas and the plight on one village**

In Padam Pokhari (Makwanpur District), LFUGs are highly frustrated and distressed with the services of Parsa Wild Life Reserves, Pathalaiya. From their efforts, their leasehold plots look like a natural forest, with more than 70 per cent crown cover. The trees they have planted are mature and ready for harvesting, as is the broom grass they have cultivated. Therefore, they need immediate thinning not only to fetch a good price from selling timber but also to slow the declining yield of forage and seeds and allow regeneration. However, despite repeated requests to the concerned authority and formal complaints to higher authorities, they have not been able to have their operational plan renewed. This has translated into a loss of thousands of rupees of income.

Source: N. Baral, during a field visit in 2011

Leasehold forestry is considered one of the most innovative and widely recognized programmes in combating poverty and rehabilitating degraded forests in the Hills of Nepal. Government policy has become progressively more supportive of it; the government is also implementing leasehold forestry from its own budget. The sustainability of leasehold forestry is very high. Despite such potential, the leasehold arrangement of those lands that have shifted under the jurisdiction of the protected area management systems is under threat (ref. 2). Some very recent efforts with communities to bring this issue to the attention of high-level authorities seem to be getting a more positive response.

# *Chapter 4*

# Impacts of leasehold forestry

The experience of leasehold forestry has shown a number of impressive results in terms of the conditions of the forests as well as the lives and livelihoods of the poorest segments of society that use them.

During the past 20 years, numerous studies and evaluations have been conducted throughout the different phases of the programme in order to understand what differences, if any, the programme was making. To measure change taking place in such a holistic and evolving approach is quite a challenge. Few programmes attempt to span improvements in tenure, natural resources management, agroforestry production, livestock development, inclusion and empowerment, microfinance, livelihoods development and well-being – all of the elements necessary to address the needs of marginal forest users. Monitoring and evaluation of such a breadth of activity can be daunting, especially when wider changes are taking place in the community.

Despite these challenges, key findings did emerge in the different dimensions of leasehold forestry outcomes (more immediate results, such as changes in forest management practice) and impacts (livelihoods and poverty reduction, environmental change and social effects). The findings come from past studies, project monitoring and evaluation reports, as well as the findings of a recent independent impact study that was commissioned by the government[7]. The independent study placed a special emphasis on comparing a robust representative and stratified random sample of

---

[7] FAO/SEEPORT, 2014. Findings of Impact Assessment of Leasehold Forestry. Kathmandu.

findings from leasehold households (groups and individual households) with those of control areas in order to understand whether the results were due to leasehold forestry efforts rather than to wider economic changes and government or other schemes. Hence the core of this chapter is based on the study unless otherwise specified.

As the impact study covers a random sample from both the first and second phases of IFAD-supported leasehold forestry projects, the results show the conservative overall estimate. In many cases, as reflected in this book, results have been much better. In addition, some notable results from the end of project report are noted from the Technical Assistance project for LFLP, which focused on four pilot districts where more intensive support was provided.

Before examining results and impacts in detail, the wider policy achievements related to the underlying legal framework for leasehold forestry should be noted.

Existing leasehold forestry is covered by the Leasehold Forestry Policy 2002 and Forest Act 1993. Other instruments that provide the policy foundations on which the leasehold forestry programme is based include: Master Plan for the Forestry Sector (1989); Agriculture Perspective Plan (1995); Tenth Five Year Plan (2002–2007); and Interim 3 Year Plans (2007–2010 and 2010–2013). All of these policies and Acts were formulated with a view that the leasehold forestry was to enhance forest product-based industries. This is a major, but somewhat incomplete, achievement. The pro-poor leasehold forestry approach was formulated when the Forest Act 1993 was already underway. Thus, the process formulated in the Forest Act 1993 is basically suited for leasehold forestry for forest-product-based industries, but much less so for the community-based leasehold forestry or pro-poor leasehold forestry. The Forest Regulation 1995 has a provision for pro-poor leasehold forestry in its Rule No. 40. However, it is incomplete, since all the other provisions for pro-poor leasehold forestry need to be borrowed from Leasehold

Forestry for Forest Product-based Industries/Ecotourism act. The Leasehold Forestry Policy (2002), which was basically formulated to resolve the ongoing issues of leasehold forestry, was approved by the Ministry of Forest and Soil Conservation but has not yet been translated into an Act and regulations. Therefore, there is still an urgent need to create a proper legal framework for pro-poor leasehold forestry to run smoothly.

A number of lower-level policy implementation mechanisms such as guidelines and manuals were prepared in order to implement leasehold forestry, and Livelihood Improvement Plan guidance has now been incorporated as an integral part of the Leasehold Operational Plans. These Livelihood Improvement Plans include:

- Guidelines on livelihood improvement planning at the household and group levels.
- Field guidelines to reactivate defunct or passive LFUGs and identify areas of support.
- Operational guidelines for integrating community forestry and leasehold forestry.
- Guidelines for integrating leasehold forestry in buffer zone and watershed management areas.
- Intra- and interdepartmental, local government and other programme coordination and collaboration mechanisms.
- Drafted constitution guidelines to federate LFUGs at the district level.

Although the agenda is not complete, the policy development process has provided a major basis for implementing the leasehold programme, and has produced some very significant results on the ground, not least the granting of leasehold tenure.

Although it may be seen as a given, probably the single most significant change that leasehold forestry has brought has been the establishment of longer-term and more inclusive tenure over forest areas by poor and marginalized forest users. The approach of the programme has been the distinguishing feature from the start. It recognises that poorer rural people often have less access to resources, and that whatever access they might have is often uncertain. The 40-year lease provides a critical source of security, as well as direct benefits to livelihoods.

The impact study showed that on average each group (about nine members per group) was handed about 5.9 hectares of forest. Almost all of the groups have Operational Plans that have been approved by the Forestry Department. In 69 per cent of the groups, LFUGs worked collectively; the remaining 31 per cent subdivided into smaller informal plots for individuals, but within overall plan agreements. In 93 per cent of the groups there was equal sharing of benefits, and for the remainder, benefits were based on needs, or priorities among the poorer group members.

As we shall see, LFUGs have been very dedicated to forestry management, and the majority of them have been sustaining their activities from the earlier leasehold forestry programmes dating back 15 to 20 years. This has resulted in a high degree of respect for the tenure agreement. The existing regulatory provisions make it possible for the Regional Forestry Directors to take back the leased forest at any time if provisions of the agreed operational plan are violated. The impact study did not reveal any cases of such violations, or cases of encroachment or capture of the leasehold forests by outsiders. These are remarkable achievements and a testament to the strengths of the policies and implementation instruments.

## Including those most in need

Complementing the issue of tenure is the programmes' special focus on the poorest forest users in the communities. As such they have also been squarely and continuously addressing the government's poverty alleviation agenda, which has been nationally recognized from the start. Marginalization has historically played an important part in conflicts, and in poverty traps.

The programmes and projects have consistently identified and worked with more marginalized groups, such as *janajatis* and *dalits*, but also poorer forest users from other castes, the food-insecure, and women, including women-headed households that often struggle to ends meet and have little voice in the community. As leasehold forestry is a commitment to poverty alleviation, it identified group members through a well-being ranking process, in which the poor were included irrespective of any ethnicity or caste (although *dalits* and ethnic households had high representation, since they are among the poorest). The impact study showed that poor people and different often marginalised social groups were at least as representative of their surrounding communities – and particularly inclusive of endangered and disadvantaged *adivasi/janajati* (74 per cent) – compared to the Village Development Committees as a whole, where they were 50 per cent of the population (in the Hill regions of Nepal, *adivasis* comprise about 34 per cent of the population). These are the groups that the latest poverty and food security studies consistently show as having the highest poverty rates, and are often traditionally dependent on forest resources. The programme has included *dalits* as beneficiaries at an increasing rate: 16 per cent compared with 8 per cent in 2005[8].

The LFUGs are generally more inclusive of poor people than the general surrounding population. The 33 per cent of LFUG households that fall below the poverty line, as determined by the Central Bureau of Statistics, was slightly higher than in control areas, and higher than the latest analysis of poverty rate of rural hill areas as whole, which is generally less than 30 per cent (except in the Far Western region, NLSS III). Further the percentage who are poor in LFUGs is higher than among self-employed farmers as a whole (27 per cent poverty rate).

These findings reveal that a combination of group and area targeting developed a considerably inclusive model for engaging the

---

[8] NPC, 2005

most marginalized and poor people who relied on forests. The impact study has also confirmed that the pattern of group inclusiveness also holds at leadership level, which signifies that decisions about resources reflect the voices of the poor. A total of 93 per cent of LFUG members have held executive positions in groups, which is twice as much compared to control areas. The activity levels of female-headed households and marginal groups, as well as their voice being heard in executive committees of local organizations, was greater in LFUG areas compared to control areas, indicating greater confidence in their roles. Moreover, *dalits* have held 13 per cent of the key positions in executive committees of LFUGs, which is little less than the overall population of 16 per cent in membership – a good example of inclusiveness.

LFUG membership includes 27 per cent female-headed households, which is a higher proportion than the population rate. These households benefit as often from the project as other households. In addition, a more general gender mainstreaming and social inclusion approach followed by the programme was found to be notably useful. Both male and female members from every beneficiary household were trained and motivated to participate more actively – 39 per cent of LFUGs key positions (chairperson, secretary, treasurer) are held by women, which is almost proportionate to the overall participation by women (42 per cent in group membership compared to only 15.4 per cent in 2006[9]. This positive change in women's position in decision-making roles has empowered women to generate income from selling forest products and not to merely harvest them. This is commensurate with previous findings[10] that "Women's participation is very high in terms of harvesting the forest products, while they have little control (25 per cent only) in selling the forest products and using the income."

---

[9] DoF, 2006
[10] NPC, 2005

Comparisons with experiences with community forest user group (CFUGs) have noted considerable challenges in the latter in terms of elite capture, and few benefits accruing to poorer households. The strength of the leasehold programmes have been their progressive inclusiveness in both membership by poorer and more marginalized ethnic groups, as well as in the level of activity by the LFUG membership of such groups.

## Managing the forests

Central to the concept of leasehold forestry is improving the condition of the forest, besides the livelihoods improvements. Bringing about change and positive impacts is challenging on two fronts. Firstly, the communities have started with very degraded lands, since this is a condition for selecting the leasehold plots, and therefore the users have to work hard to make production gains. Nevertheless evidence shows they have achieved this and have also helped to transform landscapes, thus contributing to the wider aim of environmental change. The second challenge is a more practical one: with so many small and scattered plots, it is difficult to monitor and assess the changes over a long period of time.

## Active management by poor forest user communities

Field findings from many visitors and surveys show forest user communities have transformed their landscapes as a result of tenure, capacity building and technical support. LFUGs have actively addressed the causes of degradation and managed and protected their leasehold plots to improve the condition of the forest, weeding and planting sustainable forest species that will also give longer-term household benefits. Under the impact study, household involvement in forest activities was 68 per cent compared to 22 per cent in the control areas.

The results have been less free grazing and in some cases reduced encroachment, fewer fires, and widespread planting of locally appropriate trees, grasses and herbs, including non-timber forest product species. Before the forest plots had been allocated to the user groups, free grazing and forest fires had been frequent in the vast majority of them. After allocations, they were occurring in fewer than 15 per cent of the plots. Similarly, excessive lopping of trees was also a common occurrence before, but afterward lopping was occurring in only 18 per cent of the leasehold forestry plots. (This complements the findings from the LFLP Outcome Survey showing the shift to positive management activities, such as weeding and plantation establishment (see figure below), and the shift to stall feeding for livestock.)

**Figure 8: Trends in weeding and plantation activities**

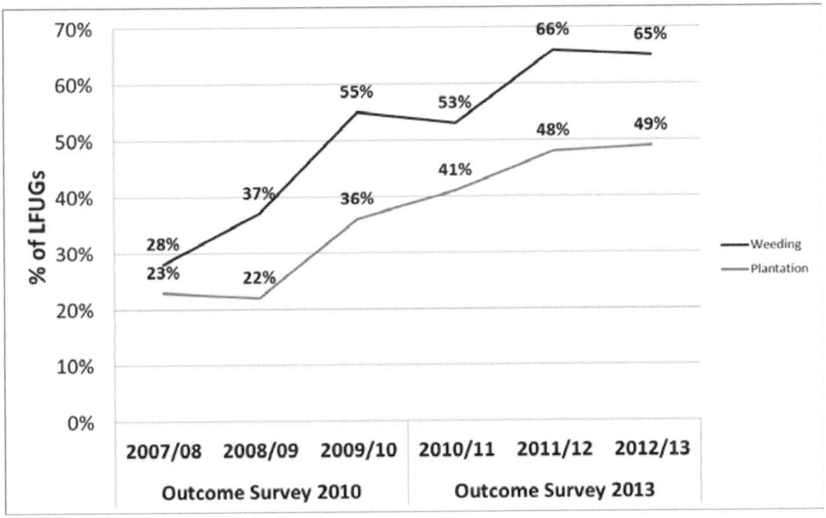


Source: LFLP Outcome Survey 2013

## Land cover change and forest improvement

While quantitative data for comparing changes in forest conditions with the original situation has not been systematically tracked in leasehold projects[11], some previous studies have examined this on a sample basis. As the leasehold plots are relatively small, it was difficult to conducts wide-scale assessments based on remote sensing. Nevertheless the impact study conducted a forest inventory which examines the changes in land cover and forest-growing stock, comparing them with available information on the baseline situation (including the Operational Plans for individual plots) and community recall. The impact study also made an effort to use comparable control forest plots and lands in order to reflect as closely as possible the conditions in which neither community forestry nor leasehold forestry have been allocated for improvement.

The findings point to substantive changes in the forest situation over the last 20 years: changes in the land cover, productivity, ecosystem health and availability of useful forest resources.

In the impact study, which is representative of all leasehold areas, more than 95 per cent of the groups have seen an improvement in the usefulness of the forest, and over 40 per cent have seen a very significant improvement, in terms of access to the forest resources, increased forest cover and availability of green fodder and forage.

The outcome surveys (covering LFLP Phase II only, see figure below) show that communities have seen considerable overall land use change across the project areas in fairly short spans of time. From 2006 to 2010 (and with similar pattern to 2013), the area under improved forest cover in leasehold plots increased by 60–70 per cent, for a majority of LFUGs, compared to the less than 20 per cent cover at the beginning, during the initial handover stage of the degraded lands.

The impact study found similar changes in productive land cover, consistent over all districts, with increases from mostly 10 per cent to over 60–70 per cent over time, which in some cases span 10–

---

[11] Critical data for the baseline situation are generally limited to the Operational Plan document, which provides a more qualitative description at the beginning.

15 years. With 51,700 hectares of leasehold area under LFLP and HLFFDP overall, this is roughly equivalent to poor and marginalized user communities having *directly* rehabilitated over 25,000–30,000 hectares of what was once bare and degraded forest land in the country.

Figure 9 shows that a majority of LFUG plots had less than 25 per cent land cover at initial handover stage (in black), compared to the survey period a few years later (in gray), where the large majority of plots had considerably improved land cover between 50 and 100 per cent.

**Figure 9: Shift in forest land cover from handover to 2010**

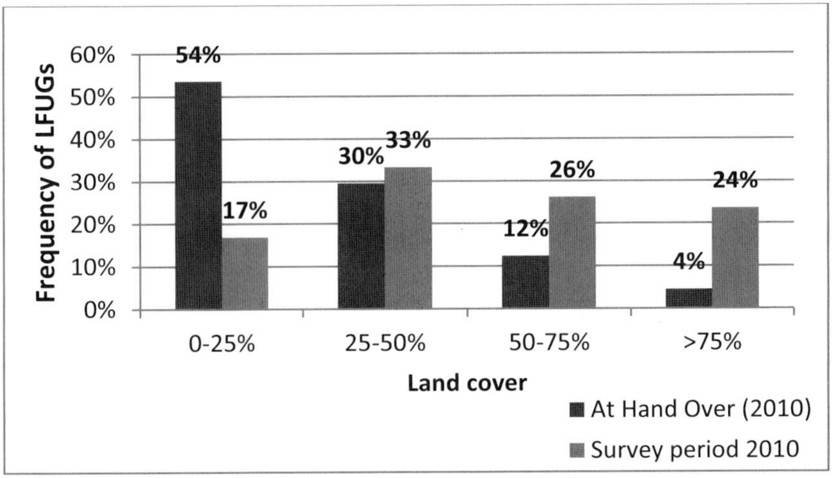


Source: LFLP Outcome Survey 2010

Although historical satellite imagery has been difficult to come by, some photographs and satellite images have shown the transformation of the land (see figure below), and in some cases the changes have taken place over just a few years. In the Technical Assistance project for LFLP, which focused on four pilot areas with a more intense landscape approach, the effects were even more dramatic. The project covered about 1,000 hectares of shifting cultivation area on the lower fragile Shiwalik hills closer to the Terai.

Before intervention in 2010, low-production shifting cultivation was performed in 52 per cent of the area, and most of the remaining forest was degraded and used for open grazing. By 2014, after a large-scale community effort to develop a broom grass/forestry-based cultivation system, broom grass and other new forestry plantations now cover the former shifting cultivation area, and all of the degraded forest is now being tended to with improved forestry practices.

What is not always obvious from photographs or remote sensing imagery is that the composition of grasses and trees has also changed[12]. The species that have regenerated and that have been planted provide more uses for the communities and play a stronger role in supporting their livelihoods. The stage has now been reached where, across all districts, between 70 and 80 per cent of the land is covered by grasses that are useful to the communities, and between 10 and 15 per cent of the land is covered by trees (see box below). In Nepal fodder is usually defined as grasses and herbs, and forage is leaves from trees which are harvested (in other countries the terms are sometimes assigned to indicate that fodder is cut and brought to the livestock, while the animals are released in the field to forage). Leasehold plots have many more valuable species compared to control plots without leasehold activities (see table 8).

---

[12] Interestingly, field observations comparing leasehold plots with neighbouring untouched degraded forest land sometimes show grassland with less vegetation cover in leasehold areas. This occurs because leasehold plots are harvested more often and are less dominated by undesirable weeds.

**Valuable fodder and forage species**

The valuable fodder species that have been grown and harvested are Mott, Badame, Mulato, Stylo and Nepier. Forage species (trees and bushes whose leaves form important feed for livestock) have also widely emerged. They include: *Litsea monopetala*, *Milea azaderach* and *M. monopetala*, *Ficus semichordata* and *F. glaberrina*, *Erythrina arborescens*. These species are being widely grown on leasehold plots in all regions. The trees that have regenerated and have been planted include: Sal (*Shorea robusta*), Asna (*Terminalia* spp.), Kimbu (*Morus alba*), Chilaune (*Schima wallichii*), Jamun (*Syzygium cuminii*) and pines (*Pinus* spp.). Non-timber forest products species include *Aloe vera* and Amala (*Phylanthus emblica*).

**Table 8: Average number of forest product species by district (range in brackets)**

| | *Fodder* | *Forage* | *Major NTFPs* |
|---|---|---|---|
| Leasehold plots | 7.0 (5–11) | 8.5 (4–12) | 12.1 (4–17) |
| Control plots | 2.5 (2–4) | 2.4 (2–3) | 3.5 (3–5) |

The forest inventory looked at 666 leasehold plots and compared them to 336 "control" plots of similar geography (i.e. with comparable slope and soil, agroecological zone, altitude) without leasehold or other forestry intervention. The comparison showed a small difference in number of trees per hectare – an average of 13/ha leasehold plots versus 10/ha in control areas (a 25 per cent difference), consistent in all districts. This was supported by 15–20 per cent more regenerating seedlings, saplings and poles. In addition, the trees in the leasehold plots have an overall average greater basal area of 26 per cent (4.38m$^2$ per hectare compared to 3.47m$^2$) and an overall greater tree volume of 28 per cent (30.1 m$^3$ per hectare versus 23.5m$^3$ per hectare. In time, it can be expected that slow-growing trees will further increase the area and volume, which means that leasehold forestry has yet to fulfil its full potential.

Other studies have shown that land cover transformation under leasehold plays a highly complementary role in the landscape and wider forestry management. For example a remote sensing study in Sindhupalchok district Gyalthum River Watershed[13] has shown how leasehold forestry activities have played a key role within a landscape of different forestry approaches over 20 years (1990–2010). This study showed that compared to other forestry regimes (community forestry, private forestry, and government forests), leasehold forestry has been the major factor in changing unproductive grassland to sparse forest, with a 300 per cent increase in the area of sparse forest. Relative to the area available, leasehold has contributed most rapidly to creating new forest areas (see figure 10 below). It has thus complemented the role of community and private forestry, which have in turn contributed to turning sparse forest into more dense forest. Community forestry and government forest have also reduced grassland, but with a greater advantage in having existing plantations to develop.

[13] Presented by Bharat K. Pokharel and Rabin R Niraula. Leasehold and Community Forestry Link: Outlook of Nepal's Leasehold Forestry, Forest Land Restoration and Poor People's Livelihoods. Page 244. Proceedings from "Leasehold Forestry: A New Dimension on Livelihoods" Regional Workshop on Pro-poor Leasehold Forestry. 11-13, June, 2014. Kathmandu, Nepal.

**Figure 10: Forest cover change in different forest management regimes**

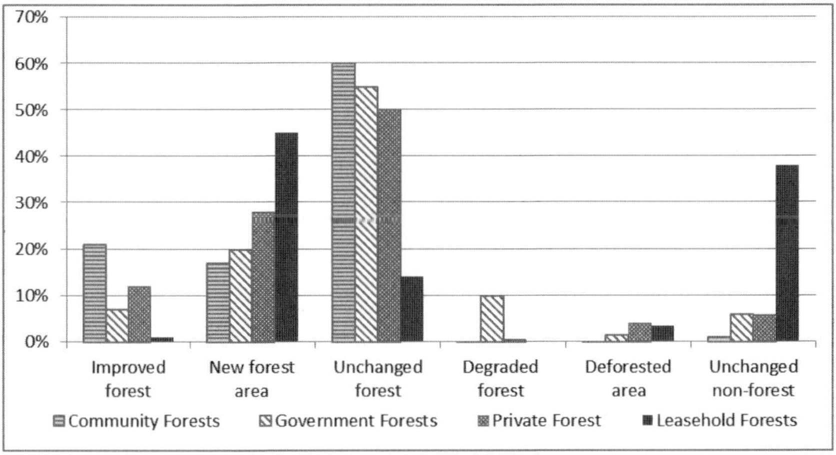


Source: Pokharel – see above paragraph footnote

As we have seen, not only has the forest cover changed, but also its usefulness has increased through the intensive activities of the households as groups. The composition of tree species has improved, and availability of forage and fodder has increased. The farmers are typically gathering more bundles (*bhari*) of fodder, forage and edible fruits from trees, closer to home and more reliably, as well as firewood and other products.

The source of forest products to meet the subsistence needs of the communities has also shifted from other areas to leasehold areas. Before leasehold forestry, community forestry areas were the most important source of fodder, forage, timber and leaf litter. After leasehold forestry, the majority of forest products are derived from leasehold plots turning the forest into a cultivated area. The only exception is timber, for which community forestry areas are still the most dominant source.

**Table 9: Sources of forest products for households to meet their subsistence needs**

| *Product* | *Fodder* | | *Forage* | | *Firewood* | | *Timber* | | *Leaf litter* | |
|---|---|---|---|---|---|---|---|---|---|---|
| *Source* | *Before* | *Now* | *Before* | *Now* | *Before* | *Now* | *Before* | *Now* | *Before* | *Now* |
| Leasehold forest | 6% | 65% | 8% | 71% | 6% | 64% | 2% | 22% | 6% | 69% |
| Community forest | 58% | 47% | 59% | 50% | 55% | 49% | 54% | 49% | 57% | 52% |
| Government forest | 12% | 6% | 11% | 7% | 16% | 7% | 10% | 4% | 12% | 8% |
| Private forest | 0% | 0% | 0% | 0% | 0% | 0% | 0% | 0% | 0% | 0% |
| Agricultural land | 5% | 2% | 7% | 4% | 5% | 3% | 3% | 3% | 5% | 4% |
| Upland (*pakho bari*) | 18% | 16% | 22% | 19% | 7% | 7% | 3% | 3% | 5% | 6% |
| Other | 63% | 59% | 64% | 61% | 56% | 46% | 40% | 32% | 43% | 33% |

Source: LFLP project.

## Livelihoods

The leasehold forestry programmes have had a specific approach: addressing the multifaceted livelihoods needs of the poorest in hilly areas – needs related to their dependency on their surrounding resources, livestock and agriculture – and creating links to other support systems as well.

## Saving time by using forest resources

Earlier studies have shown that secure access to degraded forest land, combined with training and inputs, has increased the availability of animal feed and fuel wood. The time-consuming process of gathering fodder and fuel wood is women's work, and access to leasehold forestry products reduced the time by 2.5 hours per day for each household. This in turn has allowed women to undertake more socially and economically productive activities, including learning and income-generating activities. As a result, household incomes increased, and the education and social status of women improved.

The impact study has confirmed this major effect by leasehold forestry. Firstly the more reliable availability of forage grass and tree leaves available in leasehold plots has helped to save precious time and physical effort, which is otherwise such a grinding burden for women in hill villages.

The impact study has shown that the average time required for collecting fodder and forage had been reduced from 2.6 hours to 1.4 hours, representing a 46 per cent time savings. The average time required for collecting firewood, leaf litter and timber has also been reduced, with a 20–41 per cent time saving. The precious hours regained were used in different activities, mostly for improving sanitation (49.7 per cent of households) and income-generating activities (47.6 per cent), followed by leasehold forest management (41.7 per cent), caring for old people and children (37.3 per cent), social work (33 per cent), education (28.5 per cent) and skill development (15.9 per cent). Men tended to spend more time on forest and income-generating activities, women on community and household activities. Overall, this time saving has a great influence on family livelihoods improvement.

## Livestock keeping

More than 90 per cent of rural households in project hill areas keep livestock, which are especially important for poor households and forest users who have less cropland. Livestock play a central role as a form of savings, a source of regular income, and a source of quick cash from emergency sale (medical, social) for the rural poor in the hills of Nepal. As we have seen, the key resources for livestock feeding (fodder and forage) have significantly increased in leasehold plots, and households can now rely to a great extent on their own plots.

Earlier assessments of leasehold forestry experiences have shown how increased availability of fodder made it easier to convert from free grazing to stall feeding, thus reducing the pressure on the forest and vegetation and ultimately leading to improved environmental conditions. Stall feeding also increased the availability of manure, which in turn helped maintain or improve soil fertility in the private land, leading to increased food production and food security.

Access to credit (formal and informal) encouraged households to change the composition of their livestock from local to improved animals. These animals have a higher productivity, which made it more rewarding to convert to stall feeding. As a result, more livestock products are available, contributing to improved nutritional status and food security, and increased incomes. The impact study at the end of the project in 2014 examined this in some detail.

The primary use of forest products is food for livestock, where more and higher-quality fodder and forage improves livestock health and survival, and produces larger animals and bigger yields of milk. This link between forestry and livestock has been recognized from the start of the leasehold initiatives, but in the second phase it was made explicit. While goat distribution programmes are fairly common in hills of Nepal, the impact study showed that a large majority of leasehold beneficiaries had consistently received livestock compare to only a minority in control households. The second phase of the programme included providing two goats per household, as well as strengthening veterinary support, training on animal husbandry and health, and other services (see Chapter 3).

Importantly as a result of the increased fodder and forage, animals are increasingly stall-fed, which has a significant impact on forest conservation by decreasing non-controlled grazing. Better feed and husbandry, and the promotion of local breeds, have contributed to reduced mortality and disease. The impact study showed that the use of local breeds in leasehold households was twice that of control areas. In addition 80 per cent of households reported increased kidding pattern over the last ten years, reduced mortality (81 per cent), and increased cow milk yield (57 per cent) – all better than in control areas. Only buffalo milk yields had increased for a larger per cent of control households.

The impact study also showed that the average number of she-goats (a key input from the programme) had increased by 22 per cent (from 3.78 to 4.81) compared to 7 per cent in control areas. Indeed, the herd size of goats has increased on a consistent basis, beyond the two goats originally handed out, and especially during the second phase of the programme (see figure 11 below). Here the outcome survey showed that over six to seven years, the average herd size had gone from about 4.2 to about 7.

The herd size of larger animals did not increase so significantly. However, productivity for consumption and sale – in other words, what is useful from an income and livelihoods points of view – shows interesting results when comparing leasehold areas to control areas without leasehold activities.

**Figure 11: Goat herd size (% of households)**

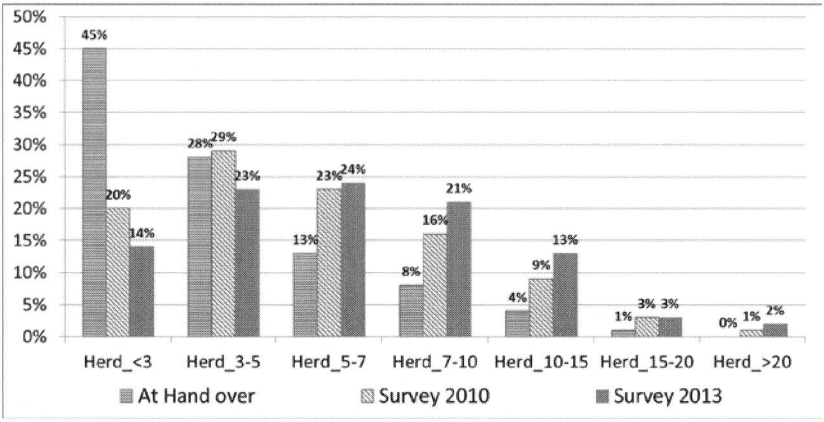


Source: LFLP Outcome survey 2013, page 11

The most important sources of income from livestock are the sale of milk and goats. Milk production was 20 per cent more than in control areas, and the number of goats sold was 11 per cent higher. This resulted in 27 per cent greater income from milk, and 12 per cent greater income from goats. The overall livestock income was 17 per cent higher in leasehold areas (average NPRs 35,800). This may seem modest, but if all production (including home consumption) is converted into Rupees using local sale prices, then the value is NPRs 97,700 (USD 977) in leasehold areas compared to NPRs 62,300 (USD 623) in control areas – a very significant 57 per cent greater value in leasehold areas. The findings also indicated that increased production has resulted in considerably greater home consumption of livestock products by leasehold households compared to control households: milk consumption is 25 per cent higher, ghee (the local butter) is 88 per cent higher, and meat a remarkable 239 per cent higher. This is important for nutrition and health in the household.

**Table 10: Comparison of livestock production, consumption, sales and incomes between leasehold households and control households**

| Product | Unit | | Production | Consumption | Sale | Income (NPRs) | Price (NPRs/ Unit) |
|---|---|---|---|---|---|---|---|
| Milk | Litre | Leasehold | 842 | 463 | 388 | 16,289 | 42 |
| | | Control | 699 | 369 | 305 | 12,800 | 42 |
| | | % difference | 20% | 25% | 27% | 27% | |
| | | | | | | | |
| Ghee | Kg | Leasehold | 6.9 | 4.5 | 2.5 | 1,088 | 428 |
| | | Control | 3.5 | 2.4 | 1.9 | 923 | 486 |
| | | % difference | 97% | 88% | 32% | 18% | |
| | | | | | | | |
| Meat | Kg | Leasehold | 20 | 16.6 | 2.6 | 1,351 | 514 |
| | | Control | 8.9 | 4.9 | 2.9 | 1,628 | 563 |
| | | % difference | 125% | 239% | −10% | −17% | |
| | | | | | | | |
| Goat | Number | Leasehold | 4.9 | 1.3 | 3.3 | 15,353 | 4,666 |
| | | Control | 4.4 | 1.5 | 2.8 | 13,686 | 4,871 |
| | | % difference | 11% | −13% | 18% | 12% | |
| | | | | | | | |
| Cow | Number | Leasehold | 0.1 | 0 | 0 | 150 | 7,500 |
| | | Control | 0.2 | 0.1 | 0 | 56 | 2,778 |
| | | % difference | | | | 170% | |
| | | | | | | | |
| Buffalo | Number | Leasehold | 0.1 | 0 | 0 | 731 | 24,353 |
| | | Control | 0.1 | 0 | 0 | 956 | 31,852 |
| | | % difference | | | | −24% | |
| | | | | | | | |
| Poultry | Number | Leasehold | 1.6 | 0.9 | 0.7 | 610 | 836 |
| | | Control | 0.7 | 0.4 | 0.3 | 238 | 767 |
| | | % difference | 129% | 125% | 133% | 157% | |
| | | | | | | | |
| **Total** | | **Leasehold** | | | | **35,772** | |
| | | **Control** | | | | **30,467** | |
| | | **% difference** | | | | **17%** | |

Source: LFLP Impact Study 2014.

## Savings and credit

Rural areas and households, especially those which are poorer, more remote and vulnerable, are characterized by having difficulty in building up savings and gaining access to appropriate credit – or by having to borrow at high interest rates which, if unpaid, leads them into binding indebtedness.

The impact study found that there was a large variation in the savings and credit activities among groups in different districts. Nevertheless, at the household level the study showed that overall the savings habit has increased among households (including female-headed households) and for all caste and ethnic groups, poorer and better off. The increase was greater in leasehold households, from 58 to 81 per cent, compared to 53 to 57 per cent in control households, especially for *janajati* households. Not only this, but the greatest relative amount saved was by the most marginalized in leasehold areas: the *dalit* and disadvantaged *janajati* households, compared to controls.

In terms of borrowing, a slightly higher percentage of leasehold households were borrowing from formal sources (75 per cent), instead of informal sources (e.g. moneylenders, neighbours) compared to control households (68 per cent). However the difference is greater with more marginal groups (*dalits* and *janajatis*) taking more loans from formal sources, indicating that the leasehold programmes seem to have helped the more marginalized groups to borrow from more reliable sources.

About half of the surveyed households took loans. The impact study revealed that, of these, 40.4 per cent of the households in the leasehold groups used loans for agriculture and livestock and 15 per cent for purchase of land and to maintain homes. Only 5.2 per cent of the households reported using loans to invest in trade/business/small industry. In the control area, only 32.4 per cent of the households reported using loans for agriculture and livestock. Overall, however, the average amount borrowed was slightly higher in control areas by *janajatis* and *dalits*.

## Incomes and expenditure

With the changes in household investments and gains, especially from livestock, it would be expected that there would also be significant changes in household incomes and expenditures. Indeed there have been substantial changes: average leasehold household incomes have increased in the last ten years by over 270 per cent, from about NPRs 39,000 (USD 390) per year to NPRs 144,000 (USD 144). However, this increase is not very different from what happened in control households. Similarly, household expenditures have increased by 143 per cent, from NPRs 47,000 (USD 470) per year to NPRs 114,000 (USD 1,140), with a similar increase in control areas. Although leasehold households have had a greater increase in income from livestock (130 per cent) compared to controls (103 per cent), it seems that such changes are small compared to the wider livelihood changes in all the hill areas. This includes the very large 10–20-fold increase in incomes from remittances, but also the 300 per cent increase in crop sales, business activities, and the like.

The question of overall livelihood benefits remains challenging, and the changes taking place within different subgroups will need to be analysed further. Gathering and analysing income data is extremely difficult, as the changes taking place within the household come from multiple income sources. For example, savings and additional livestock incomes may be used to invest in agriculture, but also to finance migration in order to receive remittances[14]. We should also note that it does not capture non-income change, such as the increased home consumption of meat and milk noted earlier.

---

[14] Appraisal on migration and remittances in Nepal. ICIMOD, IISD. Supported by IFAD, 2013.

## Food security

Earlier assessments indicated that overall household food security steadily increased for leasehold households: a family survey found a 16 per cent increase in person-months of food self-sufficiency between 1996 and 1999, compared to a 4 per cent decrease in food self-sufficiency in similar non-project households over the same period[15].

An analysis of food sufficiency data from the leasehold and control areas revealed that improvement in food sufficiency had been relatively slow for both over the decade. However, the food sufficiency level was better and improved in the leasehold area. Significantly, this was greatest for leasehold female-headed households (17–25 per cent) and endangered *janajati* groups (15–32 per cent). The reported key factors behind the relatively slow improvement were: conflict situations, lack of male members/labour in the villages for farming, lack of irrigation, and absence of improved farming practices. In the more recent Technical Assistance pilot, areas receiving more intense support that was based on many years of experience showed a higher increase. Focusing on Magar (*janajati*) communities, the increase in food-secure households (those with over 6 months of food supplies) went from 19 per cent to 50 per cent in a span of a few years. This shows that with sufficient focus and attention, poor hill communities can quickly address a fundamental livelihood concern through leasehold forestry, even while leasehold forestry is expected to provide returns for many years.

## Other livelihood and household changes

A number of past studies and field checks have shown benefits beyond production and incomes accruing in the households. Similarly, the impact study showed considerable changes and

---

[15] IFAD, LFLP Project Design Document.

improvements in sanitation, literacy, household assets, and the use of energy and tools in the last ten years – among both leasehold and control households. Outside the influence of the leasehold programmes is the fact that in control areas the transport time has been significantly reduced, due at least in part to the opening of new roads. In most of the cases leasehold groups have strengthened social links among individuals with vulnerable links, giving them trust in each other and helping them to get used to working together. This paved the way towards other economic activities and the best groups were able to grow and turn themselves into cooperatives or small enterprises.

It is also important to note that some livelihoods changes also quite subtle but have potentially long-term benefits. For example, earlier studies of leasehold showed considerable shifts in intra-household decision-making. Before the leasehold groups were formed, only 10 per cent of the women could make decisions independently, while 30 per cent made joint decisions and 60 per cent relied on a male member. After five years of project support, 25 per cent of women could decide for themselves, while 55 per cent made joint decisions and only 20 per cent relied on a male household member[16].

The programme provided support to participatory Livelihoods Improvement Plans, partly in the form of top-up funding, but perhaps more importantly as a basis for creating links with other agencies for livelihood support (such as the Department of Agriculture). Despite some setbacks, Livelihood Improvement Plans have served as an important platform to address the integrated livelihood systems of hill families and communities, with some exemplary achievements. It must be remembered that the leasehold forestry activities were taken as an "entry point" for overall livelihood development. Livelihood Improvement Plans were created for each household taking into consideration their income,

---

[16] Douglas, and Cameron 2000, Ghimire 2000 as quoted by Ohler 2000.

expenditures, skills, available resources, and the like. Similarly a group-level Livelihood Improvement Plan was prepared after the individual plans were collected. These plans included road accessibility, improving irrigation, connection to electricity, drinking water situation, etc. and then determining the best options to maximize the potential of the area. In this way, for example in Jhirubas Palpa, under the Technical Assistance pilot project, resources were able to be pooled to conduct a number of critical activities, including: a solar-powered drinking water project, road construction, collection/processing centre for broom grass, improved sanitation, solar panels for household electricity, and the introduction of green vegetables and beans.

## Coordination and collaboration

An important part of the integrated livelihoods approach was the development of coordination within the government departments and collaboration with other programmes and projects. Each district had a District Forest Coordination Committee/District Programme Coordination Committee, with guidelines in place for their functioning. This Committee was used in the programme districts to seek coordination and collaboration with all concerned government agencies, programmes, projects and development partners, including the local governments. This mechanism was also used to pool resources from all development partners. Pooling of resources was made possible by organizing a district-level coordination workshop, to which all the development partners were invited. The Livelihood Improvement Plan was presented and partners were asked if they could provide support. Whoever was ready to support, and in whatever field, was welcomed and strong follow-up took place. This process was very fruitful, along with the joint monitoring of all activities by all stakeholders. In Palpa district alone, the process resulted in pooling worth NPRs 10 million (USD 100,000) within three years.

Similarly, a Memorandum of Understanding was reached between the Department of Forests (for the programme) and the Poverty Alleviation Fund to work jointly in areas that overlapped, especially since both had the same client base and objectives. This coordination/collaboration mechanism resulted in, for example, the development of lift irrigation in Sindhupalchowk and the construction of collection centres in Sindhuli. The UNDP Microenterprise Development Programme also joined in with the programme to develop micro-entrepreneurship among programme beneficiaries. These examples point to the important role for leasehold forestry and its targeting and building capacity of key community groups.

## Environment and climate change benefits

### *Carbon sequestration and greenhouse gas emissions*

Deforestation and forest degradation contribute greatly to releasing $CO_2$ concentrations in the atmosphere. Conversely, restoring degraded forests and halting deforestation play a vital role in reducing greenhouse gas, which is important in addressing the issues of climate change. Initiatives to reduce emissions from deforestation and forest degradation (REDD+) have emerged as an additional opportunity to manage degraded forests and uplift the economic status of poor people, but more data are needed on forest carbon stocks. Some recent findings point to considerable potential, in particular in more productive areas, as well as when a landscape approach is taken.

The Technical Assistance pilot project conducted the study "Accounting of the contribution of leasehold forestry on environmental conservation and carbon trading" along with carbon sequestration studies. It also conducted collaborative research with the International Centre for Integrated Mountain Development (ICIMOD) and an expert group to study the carbon sequestration potential of leasehold forestry in two sub-watersheds of Lamjung and

Tanahun districts. In addition, postgraduate studies were conducted to assess the carbon sequestration potential of leasehold forests in one sub-watershed each in Dolakha (Charnawati) and Chitwan (Kayerkhola). The studies included detailed on-the-ground mapping and field measurements. The results of the studies showed that leasehold forests have considerable scope to enhance forest carbon stocks, but not without some limitations.

The assessment of the carbon sequestration potential of three-year-old leasehold forests in Dolakha district found the total carbon stock to be 88.77 tons per hectare. In a cluster of ten leasehold forests of the Chitwan district lying within the Kayerkhola watershed ranges (245 m to 1944 m altitude), the average carbon stock was found to be 72.49 tons per hectare[17]. The existing carbon stocks indicate that there is scope for enhancing forest carbon stocks within the leasehold forests, as current stocking is relatively low compared to community forests. However, due to high transaction and management costs for a small area of forests, leasehold forests alone may not be appropriate to claim as potential for REDD+. It may be more favourable to bundle leasehold forests with other forest management regimes so that they can benefit financially.

The results from the 2014 impact study using inventory survey data also showed that leasehold plots sequestered more carbon (ranging from 7.43 to 20.07 mt/ha) than control plots (3.3 to 16.87 mt/ha) – although overall at slightly lower levels than other studies, since the study covered both productive and non-productive areas. Thus it also showed that at the higher range, leasehold plots have considerable carbon sequestration capacity.

[17] Vaidya, Sanima (2012) Carbon Stock Estimation of Leasehold Forests in The Shaktikhor VDC of Chitwan District, Nepal. College of Applied Sciences – Nepal (Affiliated to Tribhuvan University) in the partial fulfillment of the Master of Science (M.Sc.) degree in Environmental Science of Tribhuvan University.

In 2012, the EX-ACT tool developed by FAO was used in a rapid analysis covering data from all programme areas.[18] The tool is used to provide ex-ante estimates of the impact of agriculture and forestry development projects, policies and programmes on greenhouse gas emissions and carbon sequestration, as well as financial effects and natural capital effects[19]. In this case, the analysis could make use of existing outcome (ex-post) data as well.

The EX-ACT analysis indicates that the LFLP leads to a gross effect of reduced greenhouse gas emissions and carbon sequestration, with a net greenhouse gas balance of 4,333,801 T of $CO_2$-e. This is equivalent to an average of 4.6 $TCO_2$ fixed per hectare per year and an annual project carbon balance of 206,371 $TCO_2$. Such a result shows that the reversal of land degradation has very strong benefits for climate change mitigation. The overall greenhouse gas balance translates into benefits 96 T of $CO_2$-e per hectare or 86 T of $CO_2$-e per farmer over the project term.

The study also performed a quick comparison of data from community forestry pilot sites (ICIMOD). In terms of performance per hectare, the analysis demonstrates better results for community forestry, which remained at around 9 $TCO_2$/hectare/year, while leasehold projects were below 6 $TCO_2$/hectares/year. However, it should be noted that reforestation, which is more favourable in community forestry because of better initial forest cover and easier regeneration, results in quicker sequestration per hectare. This means that leasehold forest can provide considerable benefits in the medium term, considering the initial degraded status, which even community forestry often cannot address. This points further to the concept of

---

[18] Louis Bockel and Uwe Grewer, 2012. Impact assessment of the IFAD leasehold forestry projects on climate change mitigation and natural asset stocks in Nepal (1992-2012). A project analysis using the EX-Ante C-balance Tool (EX-ACT, v.4). FAO Rome, Italy. 2012, Draft.

[19] Such as improved water, soil and watershed condition, environment services, based on standard assumptions for Nepali hill conditions. More info on EX-ACT: http://www.fao.org/tc/tcs/exact/en

leasehold and community forestry combining/complementing each other in order to generate the greatest benefits.

## Ecosystem services

The LFUG survey revealed that different environmental services had been enhanced positively after the programme[20]. A large majority of the LFUGs reported increases in: overall green vegetation (91 per cent); forest health (93 per cent); movement of birds (94 per cent); plant diversity and richness (86 per cent); improved varieties of forest species (78 per cent); increased wildlife movement (76 per cent); and increased number of trees in the farmland (78 per cent). Similarly, 45 per cent of the LFUGs noted improved control of landslips (small landslides, which destroy agricultural and forest land, as well as at times blocking roads and isolating communities).

These positive changes can be translated into more economic environmental service values as well. The impact of a project on natural capital can be incorporated into broader planning and decision-making processes that take up sustainable development and the continued provision of beneficial ecosystem services to rural and local populations. The components of natural capital usually considered are: quality of soil; increased landscape biomass; increased water availability; better crop protection/resilience from drought; incremental erosion prevented; additional flood-protected areas; incremental forestry coverage; increased biodiversity through protected areas; and incremental timber stored in forests.

On this basis, the EX-ACT analysis provided an estimate of stock variations of natural capital that the programme generated. The incremental natural capital generated was estimated at an incremental total economic value of USD 28.5 million. This is equivalent to an incremental natural capital value of USD 563 per farmer and USD 632 per hectare. This incremental value of natural capital

---

[20] Impact Study 2014.

generated by farmers is considered a public value. Added to the public value of prevented greenhouse gas emissions due to the programme intervention[21], these two public goods represent USD 2,358 per farmer.

Assessing the hypothetical option of payment for such an environmental service, which would have occurred over the 20 years of the programme (1992–2012), the equivalent of USD 116 of payments for ecosystem services would have been generated each year. Back to the present situation, such a diagnosis could be used to justify a conditional environmental payment scheme delivered over the next ten years (2013–2023), ranging up to USD 236 per year per farmer (around USD 20 per month) if 100 per cent of the public value generated would be paid back to farmers.

Such a strategy may work best where there is a more integrated landscape approach. Assessments have been conducted of the "cluster" approach (where there are larger areas of contiguous leasehold areas, rather than smaller scattered plots) in Kavre and Dhading districts[22], which have promoted allocation of large chunks of forest areas into leasehold forests, which has directly contributed to conserving sizeable sub-watershed areas. The assessment of this cluster approach explicitly found that the conservation of forest through community and leasehold forestry has increased the quantity and quality of water downstream – an important environmental service. About 15 years ago, villagers had only three dug-wells (*Kuwa*), and drinking water was scarce. The women could hardly

---

[21] The $CO_2$ fixed per farmer is valued based on the social cost of carbon (SCC), the estimated price of the damages caused by each additional ton of $CO_2$ released into the atmosphere. The SCC based on estimations of the US interagency working group (Ackerman F et al., 2010) accounts for 21 USD per t of carbon.

[22] Singh, B.K., B.R. Adhikari and H. Singh (2009) Seeing the Community and Leasehold Forestry from the Perspective of Environmental Services and its Contribution in Food Security in Nepal: A Case of Sathighar, Kavre. Paper presented in Community Forestry International Workshop in Pokhara 2009 and published in its Proceedings.

fetch 2–6 buckets (pitchers) of water a day from the dug-wells. Now they have 20 water taps, and the time it takes women to collect water has decreased from one hour to less than 5 minutes. A total of 224 households in the downstream villages have now sufficient drinking water for the whole year within a walking distance of one to five minutes, a valuable watershed environmental service enhanced by leasehold forestry.

There are also "knock-on" livelihoods benefits to be had from such environmental service. The downstream villagers have made a series of small cemented ponds/dams in the creeks to collect water at different locations to use to irrigate their vegetable plots at a commercial scale. With the increased availability of water, vegetable farming is currently being carried out on 15 hectares by 43 households. It has substantially increased their income and contributed to food security. The irrigation facility has changed the traditional crops (maize and millet crops once a year) by adding about 20 types of vegetables during the year.

The study has given a good ground to suggest that community-based forests (including leasehold forests) should bundle themselves to claim payment for the environmental services in the watersheds. This would apply mainly to hydroelectricity dams and water sources for towns and municipalities that need large quantities of clean water.

## Overall economic benefits

Few studies have been conducted to examine the financial and economic benefits of leasehold forestry. In this chapter we have seen how difficult it is to achieve across-the-board income benefits from leasehold plots, despite the considerable results derived from forestry and livestock. This is due in part to the relatively small size of plots, the changing nature of local economies and the increasing importance of remittances. Nevertheless, it is clear that substantial resource and economic benefits result when a concerted and integrated effort, rather than a piecemeal approach, is being used.

Examining the project areas, where a more holistic, community-wide and landscape approach were taken, some major economic returns have been estimated based on results from the first few years. For example, the pilot area in Palpa and Nawalparasi started as an extension of the "One Village One Product" approach. As a part of the commercialization and micro-entrepreneurship initiatives of the user group members, the pilot started planting broom grass in the barren areas of Nawalparasi and Palpa. The ultimate objective was to develop a sustainable agro-forestry system. In the very first year, the majority of the designated areas had been covered by broom grass. Benefits started accruing quickly, after which tree species were introduced. In this way the potential benefits could be harnessed with a mixture of short-, medium- and long-term strategies. The experience is described in more detail in Chapter 3. Such an agroforestry system development based on an initial fast-growing and quick-yielding broom grass seems to be providing some vital ingredients to economic upliftment. It was imperative that the design of the new intervention would be framed in such a way that at least it should surpass the regular income that the users were getting to be attractive compared to shifting cultivation, and not adversely affect the users' food security, at the same time garner enough benefit to come out of poverty as fast as possible.

A financial analysis (Govinda Kafley) of the pilots examining household benefits with and without the programme, and conducted on a relatively conservative level, indicates that over a ten-year period the internal rate of return of the project-supported economic activity could approach 60 per cent, and net a present value of NPRs 53 million (USD 530,000) – over NPRs 300,000 income per year per household (USD 3,000 per year or 250 USD per month per household, USD 50 per capita). This figure far exceeds the Nepal National Planning Commission poverty level per average household of about NPRs 100,000 (USD 1,000), even if a sizeable margin of error is taken into account. It would seem that the pilot model points to the high potential of the plantation system as a major contributor to poverty alleviation, in which all of the households involved in the endeavour would be out of poverty from the third year.

Looking at the aggregate level of leasehold programmes, to assess benefits is more challenging. Nevertheless, some relatively simple estimates can be performed to identify overall economic benefits. The study of the programme using the EX-ACT tool prepared some initial estimates.[23] Part of the financial assets generated by a project are the incremental savings and the incremental access to credit by the project beneficiaries. For the programme, incremental saving was calculated during evaluation at USD 210 per group; for the 5,042 LFLP groups, incremental savings would amount to slightly over USD 1 million.

On this basis, the total financial capital generated estimated by the study was still relatively low, at around USD 3.7 million or USD 73 per farmer overall. This incremental financial capital is not sufficient for farmers to make large leaps in livelihood development. This is in line with the changes noted by the impact study in livelihood income changes, with relatively small differences when compared to the control groups. Within this perspective, the possibility to improve financial capital with additional monetary income flows as payment for environmental services seems critical to contribute to households' efforts to climb out of poverty.

Nevertheless, taking into account additional capital invested at farm level which could be converted into cash in case of need (such as trees, livestock, material), it is possible to assess an aggregate physical capital from the programme[24]. The result shows an aggregate value of USD 9.6 million, with 16 per cent being issued

---

[23] Bockel and Grewer, 2012. Ibid.

[24] Incremental working asset at farm level (farmer equipment, water pumping, tractors, milling equipment, etc.) can be valued based on residual value at market price. The incremental value of livestock asset is accounted separately. In terms of other physical assets, there are the incremental value of owned land at their "market price".

from improved equipment and 84 per cent being issued from the incremental value of owned livestock (mostly goats). This incremental physical capital translates at farm level to an additional capital of USD 190.

Adding together natural capital, financial capital and physical capital generated by programme beneficiaries, the Ex-Act tool measures assets which strengthen landscape farming systems and household resilience to shocks, such as drought and heavy rains. The analysis showed overall improved financial and physical capital, which provides a means for farmers to face critical situations: a capital value of USD 826 per farmer and an aggregated value of USD 42 million for the whole programme area. Clearly where conditions have been more favourable climatically and from a social point of view, leasehold results are considerably greater, and sometimes outstanding.

Taken together, the findings from rigorous studies have shown that the leasehold forestry system of tenure and support, linking forest management and livelihoods, has undoubtedly resulted in positive environmental impacts and livelihoods changes, especially for the most marginalized traditional forest-user households – those who are most in need.

*Chapter 5*

# Beyond the two projects, a national perspective and institutional context

> "2011 is the last year of the Leasehold Forestry and Livestock Project. A three-year extension is being proposed, and it looks like a go. It will focus on enterprises and climate change and knowledge management for scaling-up. After those three years, perhaps we will begin a larger project. Then there is the Multi Stakeholder Forestry Programme, now in first phase. In the project design document, it is mentioned that it should cover all forestry regimes. And WUPAP in the West, of course. It has been proposed by Department of Forest and the Ministry of Forest and Soil Conservation (MFSC) to put all kinds of forestry programmes on an equal footing. The bill was already in the Ministry of Law and Justice for review, but it bounced back to MFSC. The opposition force convinced them to bounce it back. In other words, it was rejected. Only when there are elections and a draft constitution will it be possible to try to have the bill passed. The bill was submitted about two years ago."
>
> *Pashupati Nath Koirala, Forest Management Officer, Department of Forests, LFLP*

Leasehold forestry has brought about significant positive changes in the attitude and perception of local people, especially community forest user groups, local elites, government staff and development agencies, which at the outset of the initiative were totally against leasehold forestry. Leasehold forestry is being implemented within the national framework and the laws and policies which influenced forestry for the past 50 years in Nepal (they are described below). Moreover, the leasehold forestry concept and approach are being incorporated into other initiatives. The Swiss Development

Cooperation worked with leasehold forestry in Dolkha and Ramechap and also supported incorporating leasehold forestry into community forestry. Similarly, the DFID-funded Leasehold Forestry Programme developed public land forestry in line with the leasehold forestry concept. The Biodiversity Sector Programme for Siwaliks and Tarai (BISP-ST) of the government and SNV also developed various models of public land leasehold forestry in its eight Terai districts. The WUPAP project being implemented by the Ministry of Cooperatives and Poverty Reduction of the local government with support from IFAD, has basically been designed following the concept of leasehold forestry in high-altitude regions. Some of these initiatives are described in the pages that follow (ref. 2).

## Forest types and the main Acts that affect them

Forests in Nepal were never treated as a separate sector but as an integral part of other land uses that were affected by political, social, demographic and economic changes. Forest management of the country can be divided into three "eras": before 1957; 1957–1976; and 1976 until the present.

Before 1957, the forest was divided into Raikar (state-owned, mostly in terai), Kipat (community-owned, to special ethnic groups like Limbus in the east), Guthi (religious) and Birta (forest granted to private individuals who were the favourites of the ruler). Conversion of forests for agricultural purposes was encouraged. No separate forest administration existed. Likewise, no forest management system was applied. Extraction of timber for government revenue and conversion of forests into agricultural land for tax collection were the only activities performed. Most of the hill forests were privately owned.

From 1957 to 1976, several rules and regulations pertaining to forest land use and management were passed and enforced. Among them were:

a) Private Forest Nationalization Act 1957: This Act nationalized all the private forests with the hope of obtaining equitable distribution and sound management. Later this Act paved the way for community-based forest management.
b) Forest Act 1961: It was the first comprehensive legislation in the history of Nepal. This Act defined the duties of the Forest Department and prescribed penalties for forest offences. Deforestation was seen as a forest offence, but the government went on distributing forest land to landless people.
c) Forest Preservation (Special Arrangement) Act 1967: This Act enacted the Forest Preservation Special Courts under Forest Officers, defining and articulating forest offences and prescribed penalties.
d) Forest Products (sale and distribution) Rule 1970: This rule imposed a system of permits and prices even for domestic use of forest products.

Even in this period, forest was not considered as on a par with agriculture, and deforestation led to agricultural expansion.

1976 marked the beginning of human behaviour and social forestry programmes. The forests were viewed as a resource of the community and usufruct rights were granted. Community Forestry, Leasehold Forestry and Collaborative Forestry started functioning as community forest management modalities.

In recent times, two laws and related policies have had the greatest influence on forest resource tenure: the Lands Act of 1964 and the Forest Act of 1993 (which came into force in 1995).

The Lands Act of 1964 provides for ownership of land by individuals and other legally defined entities. It is designed primarily for cultivable land, and fixes land ceilings for the hills, including the mountain, Kathmandu valley (where the capital city is located) and Terai regions. However, it does not restrict landowners regarding the ways they use the land, which can include forestry purposes if the landowner chooses. Considering that farming systems in most parts of the country integrate crops and livestock, implying a need for fodder and bedding materials for livestock, the Lands Act also provides for land area in addition to cultivated land. The owner can use this "homestead land" for planting fodder and other trees and grasses (ref. 26).

The Forest Act of 1993 provides tenure systems for two types of forests: private forests and national forests. National forests are managed variously, such as through government-managed forests, protected forests and community-based forest management systems (which include leasehold and community forestry, and collaborative forest management) – while maintaining state ownership of the land. The following are the categories of forest defined by the Forest Act:

- *Government-managed forest:* National forests managed by the government. National forests are all forests other than private forest, regardless of the demarcation of their boundaries and including cultivated or uncultivated land, roads, ponds, lakes, rivers, streams and the shingly land that is surrounded by or in the vicinity of a forest.
- *Community forest:* National forests that have been entrusted to user groups (as defined in clause 25 of the Act) for development, conservation and utilization in the interest of the community.
- *Leasehold forest:* National forests that have been leased (according to clause 32 of the Act) for specified purpose(s) to a legally defined institution, forest-based industry or community.
- *Religious forest:* National forests that have been entrusted to any religious entity, group or community as specified in clause 35 of the Act.
- *Protected forest:* National forests that the government has declared protected in consideration of their environmental, scientific and/or cultural importance.
- *Private forest:* The planted or protected forests on land that belongs to an individual as per the prevailing law.

These definitions make it clear that ownership of all except private forests rests with the State. The differences among categories of forest regard only access to the forest.

Although the Forest Act created an opening for private forestry, it still reflects the Private Forest Nationalization Act of 1957 by inserting a clause (clause 39) on registration. This states that any individual or institution willing to register a private forest may do so at the District Forest Office, which can then issue a certificate of registration. The purpose of the 1957 Act, as indicated by its title, was to nationalize the then privately owned forests. Although not mandatory, the mere existence of this clause is a source of concern, especially because of the nationalization of private forests in the past.

The impact of the 1957 Act, combined with the launching of resettlement programmes, led to a decline in national forest cover, from 51 per cent in the 1950s to 45.6 per cent in 1964. To address the problem of encroachment on nationalized forests, a new Forest Act was promulgated and enforced in 1961. This was the first law specifically designed to protect national forests, while "maintaining the interest of the common people". However, this law also failed to address the problem of forest encroachment, as it declared all lands except cultivated land to be State property. Such a declaration may even have triggered the deforestation process, as the population was growing rapidly and opportunities for employment outside agriculture were not readily available.

The following table shows the repartition of forest surfaces between the five categories of national and private forest. The table shows that leasehold forestry areas are still negligible compared to community forestry areas. It also shows that a large majority of forest areas in the country are still under government management. These government-managed areas could be handed over to leasehold or community forestry groups. However it should be noted that large forest areas do not lend themselves to intensive use; and some are high-altitude areas where it might not be possible for them to be managed by community users or to be productive for leasehold users.

As it currently stands, leasehold forestry also has some restrictions in its application. For example:

1. It should not be claimed from community forestry, since community forestry gets the legal priority.
2. The crown cover of the forest should be less than 20 per cent.
3. There should be households which are below the poverty line, as defined by the National Planning Commission.
4. There should be projects which are ready to deliver in cases of leasehold forestry below the poverty line.

**Table 11: Forest area, per ownership category**

| *Category* | *Sub-category* | *Area (000 ha)* |
|---|---|---|
| National | Government-managed forest | 3,902.27 |
| National | Community forest | 1,200 |
| National | Leasehold forest | 41.73 |
| National | Collaborative forest | 57.497 |
| National | Religious forest | 0.543 |
| National | Protected forest | 711 |
| Private | Private forest | 2.3 |

Source: ref. 4 (Dept. of Forests, 16 July 2014)

## Forestry institutions – in harmony and at loggerheads

Forest institutions can essentially be placed in three categories, based on the different administrative or implementation levels: national (policy-makers and line agencies), district (programme implementation) and community (forest user groups). It is a complex system that can both advance and hinder progress.

### *National level: policy-makers and line agencies*

The Ministry of Forest and Soil Conservation, with its various departments such as the Department of Forests (the main department concerned with community and leasehold forestry), is the primary government agency that oversees forest resources in Nepal.

**Organizational structure – Ministry of Forests and Soil Conservation:**

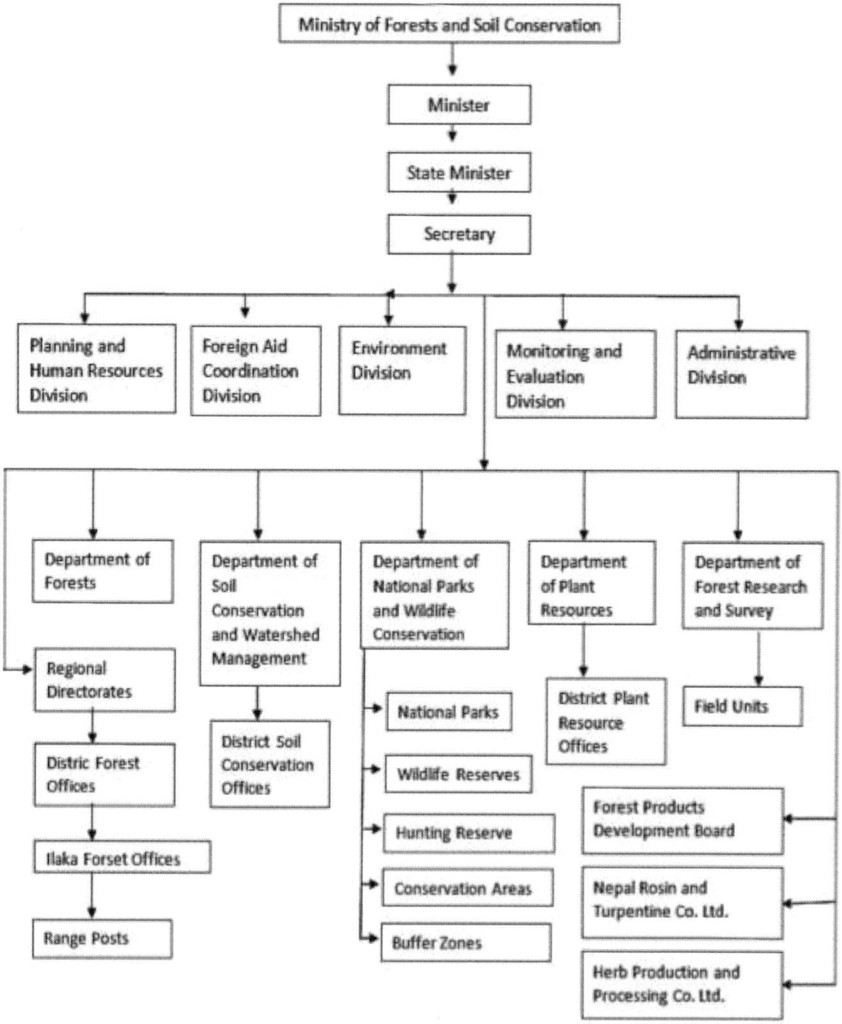


However, the Ministry of Forests and Soil Conservation is "conservative" in handing over access rights to forest users. As its name may imply, it is primarily concerned with the "conservation" of forests and soil resources, which may not necessarily favour the utilization of these resource for livelihoods, although conservation may be considered as the broader "wise use" of resources. Although the government has adopted community-based forestry management as the primary national platform for forest management, there is evidence that the national government is hesitant to hand over further rights to forest users. This is not unexpected because handing over more rights to forest communities would mean less control – and less benefit in terms of revenue, directly from forest-related taxes and royalties and indirectly through external funding from various bilateral and multilateral organizations and NGOs that have an interest in forest-related undertakings. This is paradoxical, since it was the government that introduced community-based forest management in its various forms.

Policy-makers are often influenced by interest groups that give them various types of incentives to shape policies in their favour. Those who then implement the policies (i.e. line agencies) are mostly guided by organizational incentives (e.g. perks, promotions) as they perform and accomplish outputs. However, with weak performance appraisal and incentive mechanisms, line agencies can also be prone to influence by interest groups such as local elites, traders, etc. to the disadvantage of the poorest. The primary implementing agency is the Department of Forests. The increasing popularity of integrated programmes has required the Department to partner with other line agencies (for example, with the Department of Livestock Services) and NGOs. While such partnership is ideal, given the multiple objectives of the programmes, their sectoral division within the government bureaucracy often prevents them from collaborating more effectively.

## District level – programme implementation

Policies are implemented through programmes and projects which are then implemented by the government line agencies. In the process, more specific rules and guidelines are generated that ultimately affect the access of forest users to the forest resources. These line agencies are mostly at the district level, although they are supervised and coordinated by a higher (regional/central) body. The

Department of Forests in particular, which is the primary implementing agency of these programmes, has its offices at the district level, where there are Sector Forest Offices and, under them, Area Forest Offices, both of which are administered by a District Forest Officer and Assistant Forest Officer. It is at this level where the institutions are found that guide how the forestry programme and projects are being, and should be, implemented, including the institutional arrangements between the government and the users.

The local government bodies include the nation's 75 District Development Committees and several thousand Village Development Committees, one level beneath the District Development Committees. These are under a separate government ministry – the Ministry of Federal Affairs and Local Development. Although they are not under the Ministry of Forests and Soil Conservation, the forest users, or forests, are under their administrative jurisdiction. In many instances, the District Development Committees and Village Development Committees work closely with local CFUGs, the DFOs, and along with civil society organizations to implement forestry-related and broader community development initiatives. Often there are confusions and conflicts in terms of who should be doing what.

## Community level: forest user groups

Under the various community-based forestry management programmes, forest user groups are organized and given the right to determine their own rules with regard to forest management (i.e. access and utilization), but of course under regulation by the DFO. Thus, the forest user groups, and their management committees and sub-committees, create another set of institutions. Forest user groups develop their own Constitution and Operational Plans. The Constitution basically states the rights and responsibilities of the members. The Operational Plan guides the members towards the sustainable management (i.e. protection and utilization) of the forest resources.

Forest user groups are supposed to be the main stakeholders of forestry in Nepal, considering that it is their lives and livelihoods that are directly at stake and because the programmes will not succeed without their cooperation. However, their role in the shaping of forest institutions has been minimal. Despite the enactment of the Forest Act of 1993 and its Forest Regulations of 1995, which institutionalized community-based forest management and put the users as the main stakeholders, many succeeding directives and amendments have been contradictory to the ideals of community-based forest management, and sometimes subverting positive changes through a non-participatory, top–down process.

Within the forest user groups themselves, there is disparity in interest. This can be attributed in great part to the highly differentiated Nepali society and caste system, which creates or reinforces institutions that exclude those at the lowest end – the poorest and marginalized groups – especially in rural areas (ref. 1).

This is in this context that leasehold forestry appeared in the 80's and is now supported by a broad range of projects and programmes as described in the coming pages.

## Taking the leasehold concept to higher altitudes: the Western Upland Poverty Alleviation Project (WUPAP)

The Western Uplands of Nepal are characterized by their remoteness, harsh terrain and low rainfall, which make access to and within the area poor and severely limit opportunities for agriculture. In fact the area has the smallest agricultural landholdings in the country (0.5 hectares). Those who live in these mountain areas are subject to food insecurity. Chronic malnutrition rates among young children run between 60 and 86 per cent, one of the highest recorded rates in the world. Furthermore, the area harbours the largest proportion of *dalits* in the country. Service delivery to these upland areas is very difficult, due to lack of infrastructure and understaffing of line agencies. Up to the 2010s, there has been little or no donor assistance to the area and relatively few programmes have been undertaken to assist the poor.

The remoteness of the project area, its potential for livestock and forestry production, the concentration of the poor and their lack of access to resources all pointed to the need for an intervention to assist them, as well as to stave off political insurgency. IFAD started to support development programmes in the area in the 1990s.

The Western Upland Poverty Alleviation Project (WUPAP) has been running in the western region of Nepal in 11 districts since January 2003. The IFAD-funded project entered its third phase in July 2012 and is expected to continue until July 2016. The WUPAP had five components during first and second phases. The forestry activities are implemented in a specific sub-component by District Forest Offices and follow similar leasehold handover processes to the Leasehold Forestry and Livestock Programme. The Leasehold Forest and Non-Timber Forest Product (NTFP) component is the main thrust of the project in high-altitude areas.

The project area covers 11 upland Districts in the far and mid-western development regions, with approximately 226,000 households and a population of approximately 1.2 million. By the end of the project, some 115,000 households are expected to have benefited. Priority is given to the most disadvantaged members of the community (women, *dalits*, landless or semi-landless households and other minority groups). However, in view of the overall poverty of the project area, the project has adopted an inclusive approach so that all community members will benefit.

## Components

### *Infrastructure development*

Infrastructure facilities in the project area are grossly inadequate, and certainly beyond the scope of the project to address in their entirety. However, building on recent infrastructure work by the government, the project has set up a "green road" link (corridor approach) to

# CHAPTER FOUR
## Impacts of leasehold forestry

**27.** Annapurna Dahal, an IFAD beneficiary, plucking tomatoes in her garden. She sells her produce to families in her community in Naala, as well as at the marketplace in central Banepa, Kavre.

**28.** Broom grass plantation at Jhirubas Leasehold Forest in Palpa District. Broom grass is ideal in leasehold forestry – the roots hold the soil tight to prevent erosion, the straws are made into broomsticks, and the rest is used as nutritious livestock feed.

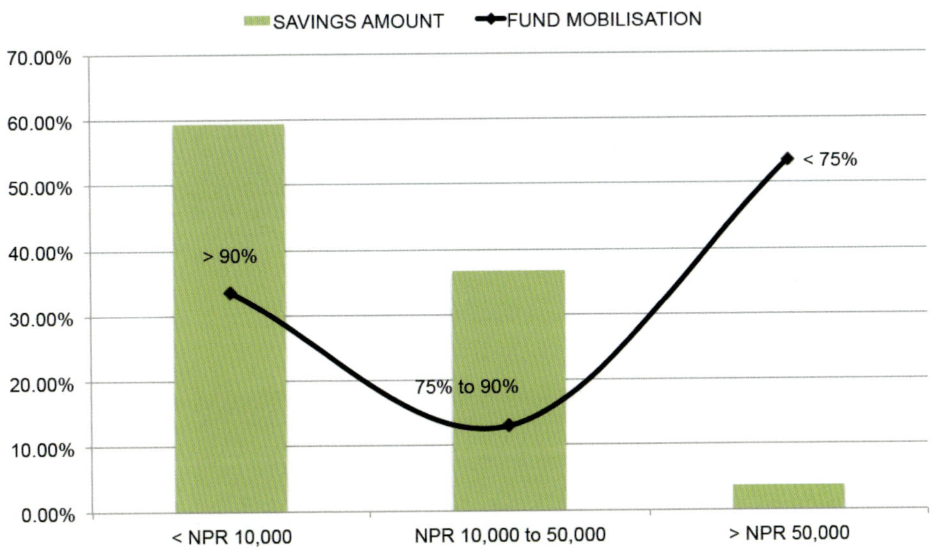


**29.** LFLP group savings. The multiple hours of time saved through leasehold forestry, in addition to financial support from group savings, has allowed LFUG members to invest in various income-generating activities, including vegetable farming, broom making and textiles.

**30.** Changes in landscape in Makwanpur District 1993–2009.

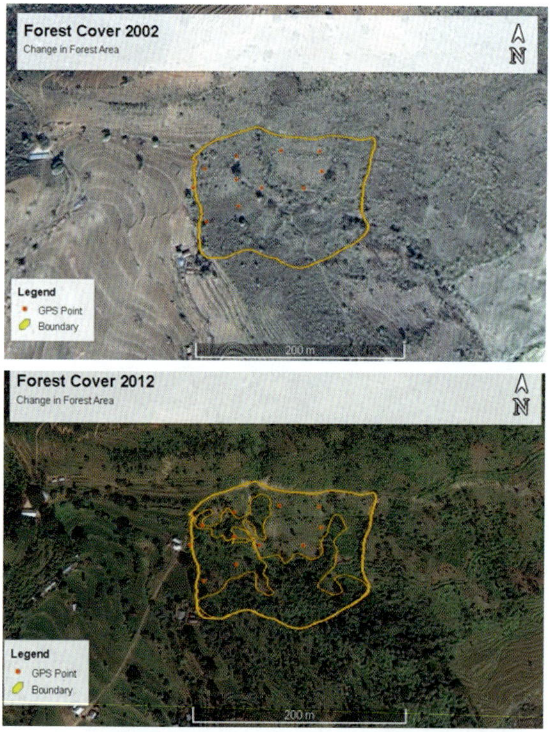


**31.** Changes in landscape in Gorkha district.

2004

2012

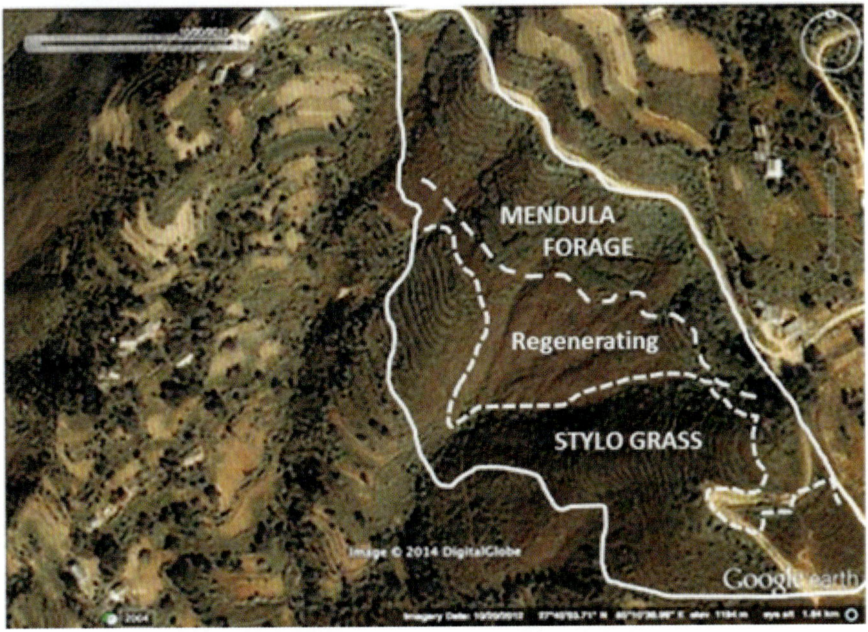


**32.** Changes in landscape in Jeewanpur LFUG, Dhading District.

**33.** Broom grass plantations supported by the Technical Assistance project in Palpa, Shiwalik Hills.

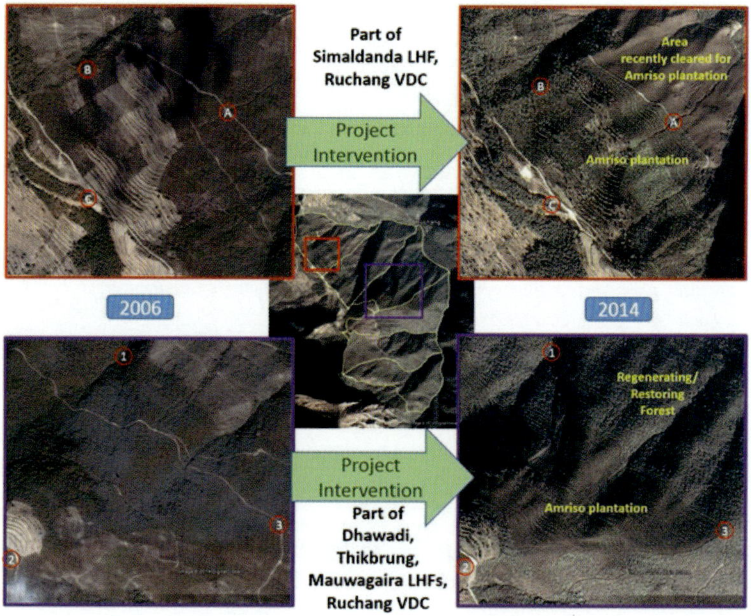


**34.** Changes in land use in two leasehold plots in the Technical Assistance project area in Ruchang.

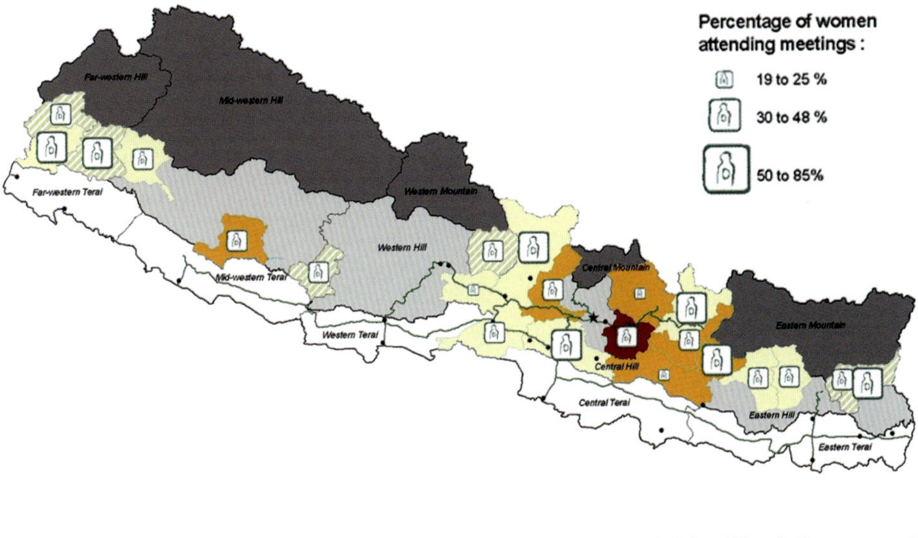


**35.** LFLP Policy dialogue Saraswoti Makwanpur LFUG. (Top) Saraswoti Tamang, Chairperson of Ramanthali LFUG in Makwanpur, sits amidst national policy makers at the Regional Workshop on Pro-Poor LF in Kathmandu discussing leasehold forestry policies. (Bottom) In the leasehold forests, at group meetings, and at trainings and workshops, the active participation and leadership of women is clearly visible in the programme.

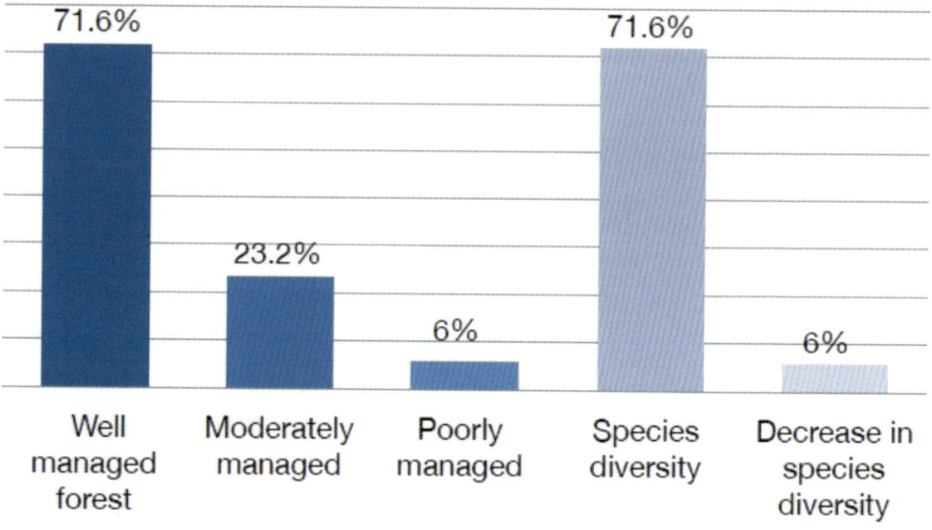


**36.** A workshop was organized on the 14th of March 2013 to celebrate the successes of LFLP, and to discuss new directions for the leasehold forestry programme in Nepal. With up to 90% improvement in ground cover in 8 years, the leasehold forestry programme has had notable positive impact on species diversity. In addition to new types of vegetation, LFUG members also note an increased incidence of birds and animals within the leasehold forest area.

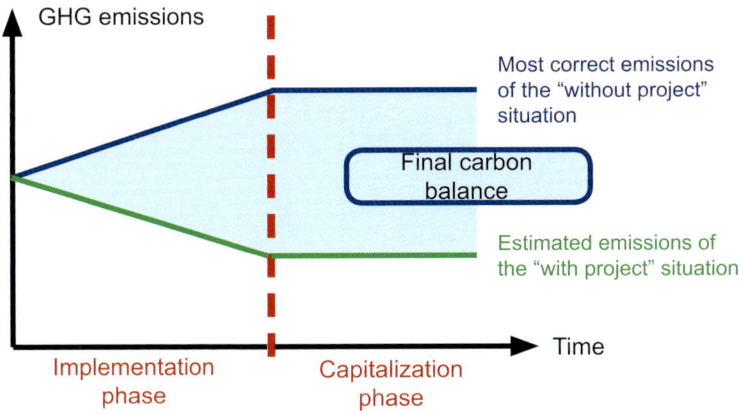


**37.** LFLP Workshop: FAO environmental economist conducts a training on using the EX-ACT tool. Deeply rooted on the rehabilitation of degraded forestland, the leasehold forestry programme has great potential of not just supporting biodiversity, but of contributing towards the management of greenhouse gases as well.

# CHAPTER FIVE

## Beyond the two projects, a national perspective

**38.** Rural village with community and leasehold forests.

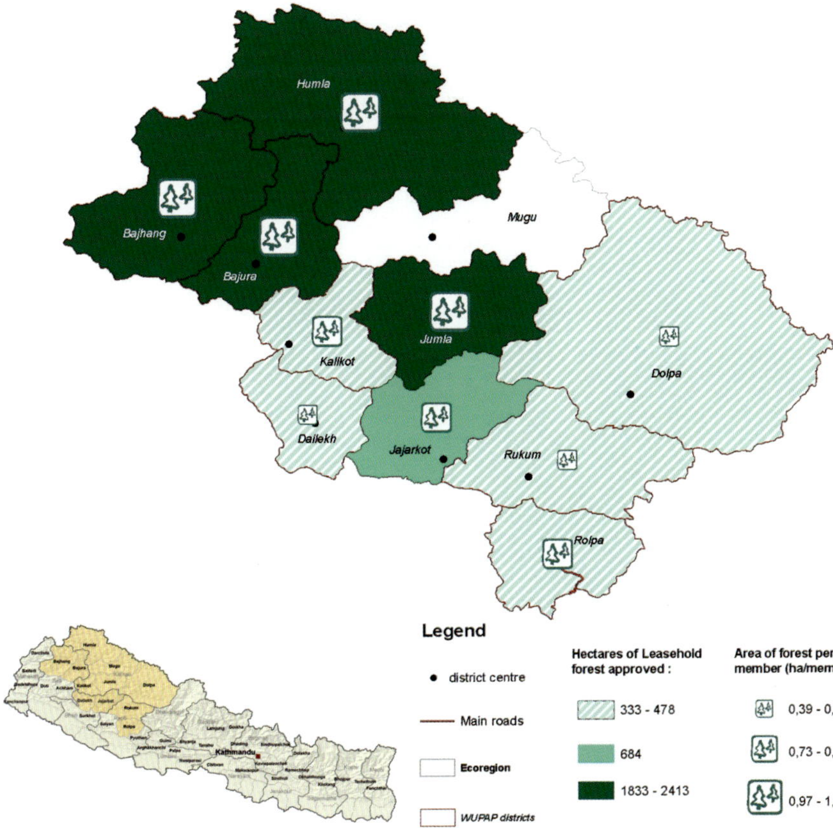


**39.** WUPAP support to leasehold activities was implemented in ten south west districts.

**40.** LFLP Kavre: a member of Chisapani Puchaar Leasehold Forest Group returning home after collecting firewood from the leasehold forest.

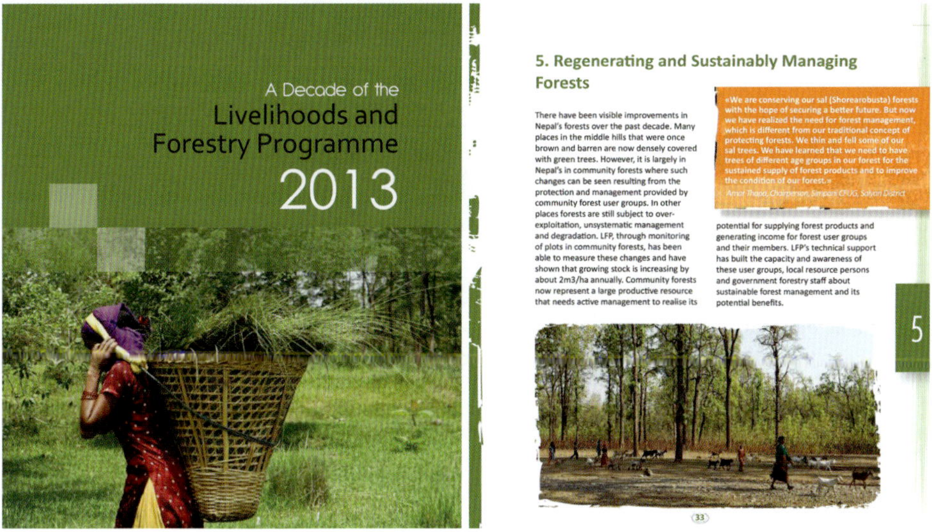
A Decade of the
Livelihoods and
Forestry Programme
2013

5. Regenerating and Sustainably Managing Forests

There have been visible improvements in Nepal's forests over the past decade. Many places in the middle hills that were once brown and barren are now densely covered with green trees. However, it is largely in Nepal's in community forests where such changes can be seen resulting from the protection and management provided by community forest user groups. In other places forests are still subject to over-exploitation, unsystematic management and degradation. LFP, through monitoring of plots in community forests, has been able to measure these changes and have shown that growing stock is increasing by about 2m3/ha annually. Community forests now represent a large productive resource that needs active management to realise its potential for supplying forest products and generating income for forest user groups and their members. LFP's technical support has built the capacity and awareness of these user groups, local resource persons and government forestry staff about sustainable forest management and its potential benefits.

«We are conserving our sal (Shorearobusta) forests with the hope of securing a better future. But now we have realized the need for forest management, which is different from our traditional concept of protecting forests. We thin and fell some of our sal trees. We have learned that we need to have trees of different age groups in our forest for the sustained supply of forest products and to improve the condition of our forest.»

*Amar Thapa, Chairperson, Simpani CFUG, Salyan District*

5

33

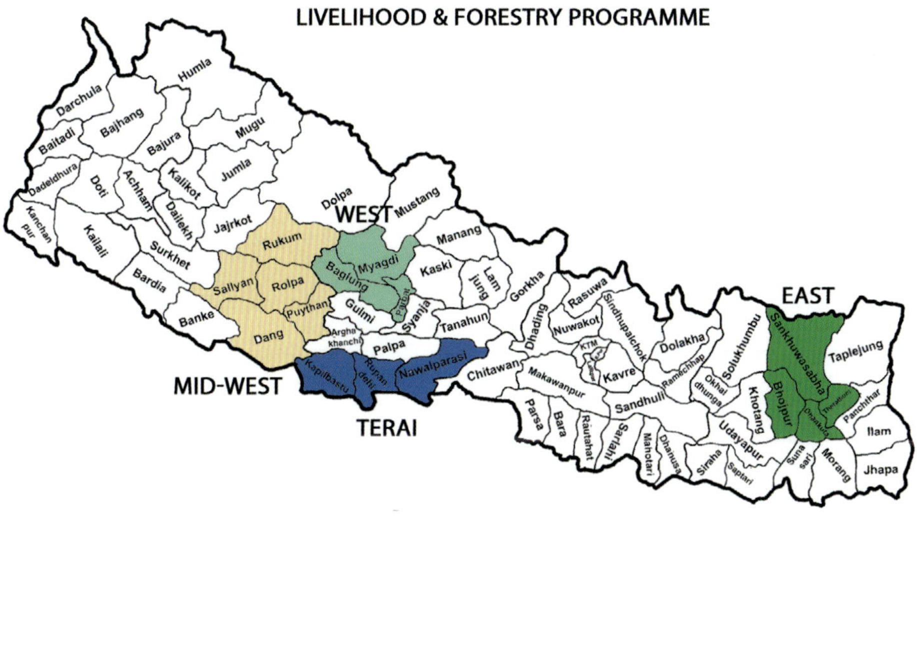


**41.** In recent years, other programmes and publications in support of leasehold forestry have emerged.

24 Decentralised Forest Sector Governance

Chapter 3

## Contribution of BISEP-ST on Nepal's Decentralised Forestry Sector Governance:

**Best Practices for Up-scaling**

BISEP-ST had several activities related to forest governance. These were implemented with its operational plan. BISEP-ST had very clear and coherent goal about the enhancing of the livelihoods of the forest dependent poor and marginalised families through strengthening and scaling up the forest governance at field to central level. BISEP-ST has been successful to achieve its general goals even in the difficult political situation of the country. The programme was successfully implemented and has had good effects or impact on sustainable forest management in eight programme districts of the programme.

During this programme period, BISEP-ST carried out several studies like reviews, student research, performance evaluation, consolidation and impact evaluation through several stakeholders. From the findings, facts and discussion, the programme has found that forestry sector has good improvement on democratisation with good forest governance, which has significantly contributed for the sustainable forest management in the programme districts. There are remarkable results and case studies which indicate that tremendous improvements have been achieved in the forest governance that ensures transparency, accountability, participation, institutional development, quality improvements and increased access to benefit sharing mechanism of poor and disadvantaged forest dependents.

*a) Decentralised Forest Management for Good Forest Governance*

Forest condition is improved (Consolidation mission, 2007). According to the report and impact study 2009, it has been seen that deforestation/degradation has been controlled and the situation has slowed, however, actual data is lacking. No hard and fast analysis has been done yet, comparing baseline data with today's situation. There are some good examples and credible claims on good forest management (such as in

**42.** BISEP supports forestry decentralized policy.

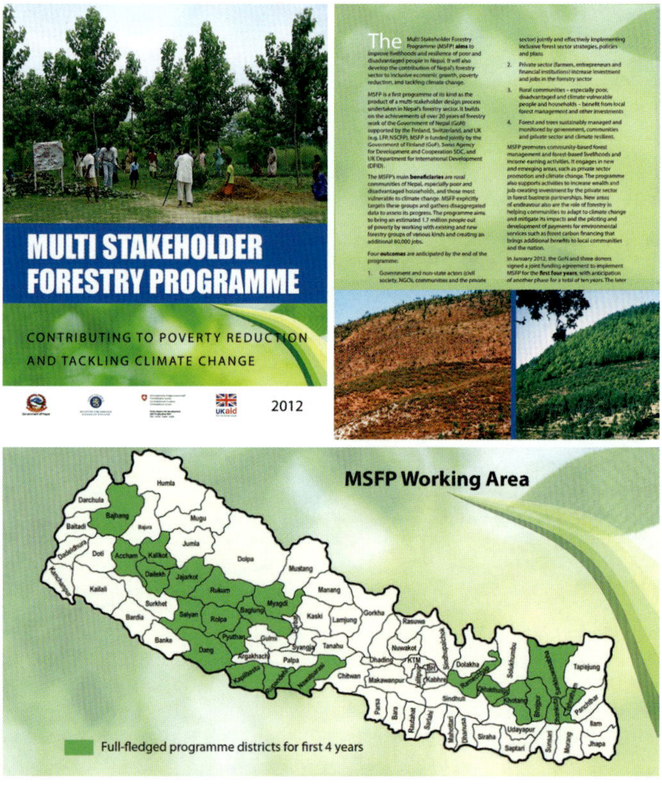

MULTI STAKEHOLDER FORESTRY PROGRAMME

CONTRIBUTING TO POVERTY REDUCTION AND TACKLING CLIMATE CHANGE

2012

The Multi Stakeholder Forestry Programme (MSFP) **aims** to improve livelihoods and resilience of poor and disadvantaged people in Nepal. It will also develop the contribution of Nepal's forestry sector to inclusive economic growth, poverty reduction, and tackling climate change.

MSFP is a first programme of its kind as the product of a multi-stakeholder design process undertaken in Nepal's forestry sector. It builds on the achievements of over 20 years of forestry work of the Government of Nepal (GoN) supported by the Finland, Switzerland, and UK (e.g. LFP, NSCFP). MSFP is funded jointly by the Government of Finland (GoF), Swiss Agency for Development and Cooperation SDC, and UK Department for International Development (DFID).

The MSFP's main **beneficiaries** are rural communities of Nepal, especially poor and disadvantaged households, and those most vulnerable to climate change. MSFP explicitly targets these groups and gathers disaggregated data to assess its progress. The programme aims to bring an estimated 1.7 million people out of poverty by working with existing and new forestry groups of various kinds and creating an additional 80,000 jobs.

Four **outcomes** are anticipated by the end of the programme:

1. Government and non-state actors (civil society, NGOs, communities and the private sector) jointly and effectively implementing inclusive forest sector strategies, policies and plans
2. Private sector (farmers, entrepreneurs and financial institutions) increase investment and jobs in the forestry sector
3. Rural communities – especially poor, disadvantaged and climate vulnerable people and households – benefit from local forest management and other investments
4. Forest and trees sustainably managed and monitored by government, communities and private sector and climate resilient.

MSFP promotes community-based forest management and forest-based livelihoods and income earning activities. It engages in new and emerging areas, such as private sector promotion and climate change. The programme also supports activities to increase wealth and job-creating investment by the private sector in forest business partnerships. New areas of endeavour also are the role of forestry in helping communities to adapt to climate change and mitigate its impacts and the piloting and development of payments for environmental services such as forest carbon financing that brings additional benefits to local communities and the nation.

In January 2012, the GoN and three donors signed a joint funding agreement to implement MSFP for the **first four years**, with anticipation of another phase for a total of ten years. The later



**43.** MSFP is a multi-stakeholder funded programme aiming at long-term activities.

# CHAPTER SIX

## When the programme leaves

LEASEHOLD FORESTRY IN NEPAL

OVER 20 YEARS OF LEASEHOLD FORESTRY IN NEPAL

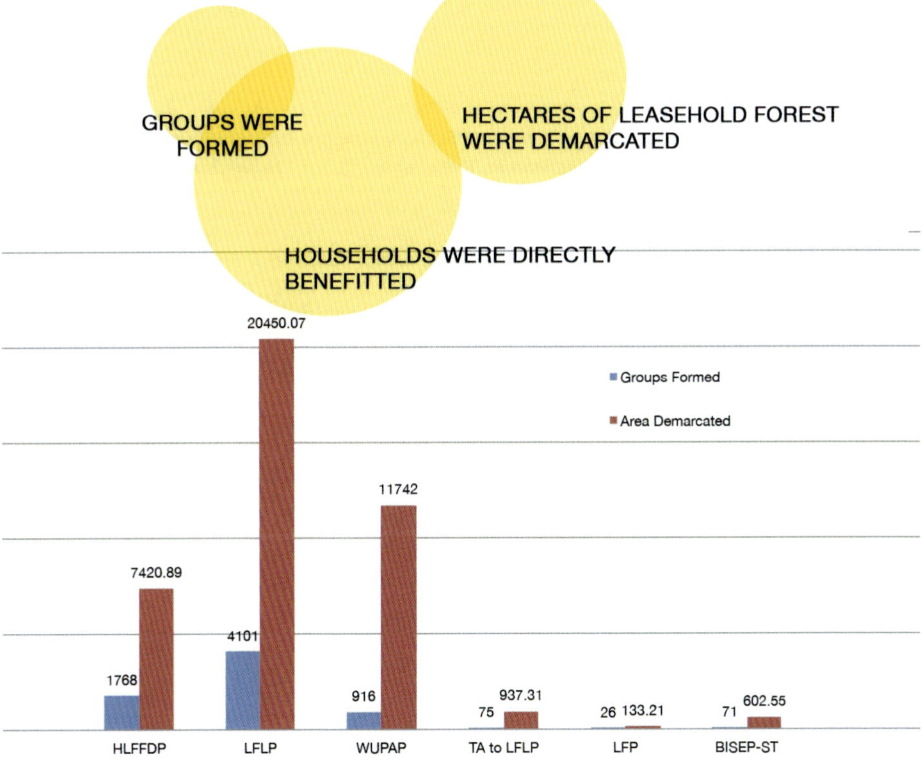


**44.** The overall results and impacts of LFLP in Nepal.

Unofficial Translation

Government of Nepal
Ministry of Forest and Soil Conservation
Singhdarbar, Kathmandu

11 December
International MOUNTAIN DAY

## Mountain Development Award

On the auspicious occasion of *International Mountain Development Day*,

Recognizing the remarkable contribution for Sustainable Development of Mountains of Nepal

**Kauledanda Leasehold Forest Intergroup**

**Jhirubas, Palpa**

*has been awarded* Mountain Development Award - 2010 *with this certificate and a cash prize of NPR 200,000/-*

| | |
|---|---|
| Yuvraj Bhusal<br>Secretary<br>Ministry of Forests andSoil Conservation | Dipak Bohara<br>Minister<br>Ministry of Forests and Soil Conservation |

Date: Paush 16, 2067 B.S. (December 31, 2010)

**45.** Many LFLP groups received national or international awards for environmental protection throughout the past 20 years.

provide access from the remote northwestern districts to the *terai*. There is little doubt that such a road network has made upland production economically viable. It has enabled food to be delivered to chronically food-deficient areas. The project is also developing demand-driven, small-scale infrastructure, which includes irrigation, foot and mule trails, bridges, water supply and storage facilities, which the communities will maintain.

### *Leasehold forestry and NTFP production*

Building on lessons learned from the Hills Leasehold Forestry and Forage Development Project and Leasehold Forestry and Livestock Programme, this component is promoting the transfer of lands, on a leasehold basis, to women and the landless to enable them to: (i) undertake leasehold forestry; (ii) increase their production of fodder crops (with emphasis on the production of forage seed as an income-generating activity); and (iii) undertake NTFP production. The project aims to make 22,500 hectares of marginal lands available to 1,000 LFUGs, comprising 15,000 people.

For NTFP production, the project is introducing domesticated NTFPs that are socio-economically viable and meet market demand. The private sector is involved in this component (including the firms Dabur Nepal Ltd and Herbal Processing and Production Ltd) in terms of identifying suitable areas for production, providing quality saplings, transferring technology and purchasing produce. In addition, one nursery is being set up for every ten LFUGs in the project area, as well as one at each District headquarters as holding nurseries to serve as demonstration sites. Women are being trained to operate the nurseries and to provide technical advice to the beneficiaries (see box below).

### *Crop and livestock production*

In addition to promoting increased production of small stock and staple food crops (new varieties of rice, maize, wheat and potatoes), this component is promoting sustainable production practices –

composting, proper husbandry, integration of fodder crops, introduction of legumes such as black grams and soya beans, increased livestock production through better nutrition and improved animal health. It is also promoting diversified kitchen gardens, providing nutrition training and organizing village-based monitoring of malnutrition.

### *Rural microfinance*

This component is working with village-based community organizations to deliver savings and credit services. Savings are the cornerstone of the credit programme. After being trained, community organizations will be required to generate and manage savings and lending from their own sources for at least six months before they become eligible to borrow project funds from the Local Development Funds. The project will grant these funds only once community organization managers have demonstrated their ability to: (i) manage the operations; (ii) maintain high recovery rates from the savings funds; and (iii) keep proper records. Loans will be granted for all viable activities, including the production and marketing of NTFPs; agricultural and livestock production; and off-farm activities, including food processing, trading and marketing. The poorest members of the community organizations will be granted loans on a priority basis.

### *Institutional support*

Given the limitations in line agency staffing and in communications in remote areas, the development of a strong institutional framework is critical. To this end, this component is financing a number of initiatives, including: (i) training to build local capacity and leadership; (ii) the cost of establishing the Project Coordination Unit in Nepalganj and District offices; (iii) the cost of any technical support required; (iv) the contracting NGOs or other service providers; and (v) development of the project monitoring and evaluation system (ref. 11).

**Cultivating non-timber forest products in high altitudes of Western Nepal**

Non-timber forest products (NTFPs), particularly medicinal and aromatic plants (MAPs), have enormous potential as an income-generating option for poor smallholders living in high altitudes of Western Nepal. Nepal is already competitive in the sector, but raw materials are currently collected from the wild, often in intensely unsustainable ways.

The Western Uplands Poverty Alleviation Project piloted demonstrations on the domestication and promotion of NTFPs. IFAD also provided grant resources to the International Centre for Integrated Mountain Development (ICIMOD) to implement the programme "Securing Livelihoods in Uplands and Mountains of the Hindu Kush-Himalayas, 2000-2005". The programme focused on growing medicinal plants in leasehold plots to supplement income generation.

As a first step, the programme conducted performance trials to select appropriate species. A private pharmaceutical firm – Dabur-Nepal – participated in the selection process, and two pilot demonstration sites and seven species were selected for the trials. Of these seven species, Dabur-Nepal recommended five species to promote and cultivate in two sites.

After the performance trials were completed, the project had to build the capacity of LFUG members, personnel of the line department and members of the project team to promote and sustain domestication. ICIMOD provided a technical consultant to conduct training in the concept of leasehold forestry, development of leased land, establishment and management of nurseries, and sustainable harvesting and management of NTFPs. Fifty-five people were trained, including 15 women.

These efforts led to the following positive results:

- The experimental trials pinpointed potential species that could be grown in different locations.
- The trials demonstrated an institutional model for fostering the participation of the private sector in livelihood security initiatives – a model which can be replicated.
- Out of the 12 nursery operators the project trained, two are involved in operating special nurseries run by the District Forest Office. These nurseries produce seedlings at low temperatures and harden them before they are transported.
- The other ten operators became responsible for operating village nurseries.

Pilot demonstrations for domestication of MAPs were initiated on leasehold plots in two districts in April 2006. Encouraging results led to the programme scaling up the demonstrations to two more districts in 2007. The scaling up also resulted in a diversification of domesticated species.

To promote NTFP cultivation, it is fundamental to identify potential leasehold plots that have a reliable source for irrigation nearby. The Western Uplands Poverty Alleviation Project identified such plots – 199 hectares in Jumla and 192 hectares in Humla.

In Jumla, the project has prepared irrigation schemes to cover 41 hectares of leasehold plots and brought 5.55 hectares under NTFP domestication. In Humla, the project has managed irrigation for 72 hectares of leasehold plots and 5 hectares have already been brought under NTFP domestication. By July 2007, a total of 19.45 hectares of lease land had been brought under NTFP domestication, benefiting 373 households. The project plans to scale up the domestication of NTFPs to 47 hectares of leasehold plots. To support the expansion efforts, the project has planned 22 irrigation schemes for lease land, to be completed by mid-July 2008. The project also plans to cover new districts for piloting demonstrations and to introduce the Sloping Agriculture Land Technology.

As efforts for scaling up progressed, the programme felt the need to forge more effective partnerships with the private sector.

ICIMOD initiated discussions with private firms operating in the NTFP and MAP sector, and has initiated an agreement with Male International. Once signed, Male International will provide technical support and capacity building to project LFUGs, and will market certified organic MAPs from project districts.

In addition to the leasehold plots recommended by Dabur-Nepal, NTFPs were occurring naturally in leasehold plots of other districts, which need to be surveyed – in particular over 5,200 hectares of land handed over to LFUGs consisting of poor households (especially woman-headed), *dalits* and landless people under WUPAP in Bajhang, Bajura, Humla and Jumla.

Interactions with LFUG members reveal that many of them have in-depth knowledge about different aspects of naturally occurring NTFPs. They could provide training to other members and should also participate in promotional and scaling-up exercises in the future. Their efforts should be complemented by the involvement of traditional medicinal practitioners.

Marketing strategies are important. Participation of the private sector alone is not sufficient to promote the NTFP sector. LFUGs should be organized into federations for collective marketing. To make attract private-sector partnership and increase LFUG bargaining power, it is necessary for products to be collected at designated collection centres to ensure economically viable volumes.

There is a risk that dependency on the private sector for the selection of species may result in species that do not have attributes desirable by the community. For example, experience from the piloting has shown that species recommended by the private-sector partner may have a long gestation period. There is also uncertainty about the quality of the final product. Dependency may also minimize diversification, which is an important risk aversion strategy.

*Sanjeev Kumar Shrestha, NTFP & Marketing Specialist, Western Upland Poverty Alleviation Project, Nepal*

*Bashu Babu Aryal, Field Presence Officer, Nepal*
*Eklabya Sharma, Programme Manager and Dhrupad Choudhury, Programme Coordinator, ICIMOD, Kathmandu, Nepal*

Source: Making a Difference in Asia and the Pacific, Issue 18, December 2007.

## Achievements to date

- More than 12,000 households have leased 11,326 hectares of forest land through 887 groups formed in ten districts.
- Horticultural plants and Medicinal and Aromatic Plants (MAPs) have been cultivated on the leasehold plots. They include mangos, oranges, *Amala*, *Allo*, *Dahatelo* and *Kutki*.
- Seabuck thorn fruit plants (*Hippophae salicifolia* and *tibetana*) have been domesticated. The plants are grown along river belts in high altitudes and produce quality vitamin-enriched juice. Processing units have also been established in several districts.
- Nurseries have been set up in leasehold plots in all of the districts.
- About one third of degraded forest land has been regenerated and rehabilitated by increasing forest and forage coverage.

At the same time, many of the leasehold plots are not properly managed because of their distance from the residence/community of the lessee. Remoteness is also a challenge for project staff and technicians to provide their much needed services. Many are reluctant to travel frequently or to spend a lot of time in the districts' hills because of the remoteness and harshness. In addition, open grazing is still taking place in about 20 per cent of the area.

It is clear that high-altitude communities require more time, effort, inputs and funding – to cultivate their products, to receive technical services on a regular and long-term basis, and to market any surplus. In this regard, the local resource person will play a major role, serving as a model entrepreneur in NTFP product processing and in other forestry-related income-generating activities. This could go a long way in helping groups to process and market the MAPs and fruit trees that are producing significant yields and offering the hope of better livelihoods (ref. 23).

## The Livelihoods and Forestry Programme (2001–2011)

The Livelihoods and Forestry Programme (LFP) was a ten-year programme that began in 2001. Building on the experiences of the Nepal-UK Community Forestry Project, the programme worked to reduce poverty and vulnerability, especially among poor and excluded people. LFP focused on enhancing household and community assets – especially by promoting more equitable and sustainable use of the forests and other natural resources on which poor rural people depend. The project was funded by the UK's Department for International Development (DFID) through a bilateral arrangement with the Government of Nepal's Ministry of Forests and Soil Conservation. Working with a wide range of government and civil society partners, the programme reached over 4,500 forest user groups covering 527,000 households (13 per cent of Nepal's population).

Of the households covered by LFP, about 53 per cent were identified through a participatory well-being assessment as being "poor" or "very poor". LFP partners work specifically with interest groups within the user groups that represent these households to identify and address their particular needs.

LFP-supported activities included: building the capacity of forest users, forest managers and service providers to manage natural resources equitable and sustainable; encouraging livelihood diversification and income-generating activities for poor and excluded households; and developing enterprise and small-scale infrastructure.

The programme was implemented in four very different areas of the country: the Koshi hills in the East; the Dhaulagiri hills and Lumbini plains (Terai) in the West; and the Rapti hills in the Midwest. In the plains, LFP worked with groups connected with community, leasehold and national forests, as well as public land.

LFP created a Pro-Poor and Social Inclusion (PPSI) Strategy, as well as a social mobilization programme, to address social exclusion – whether based on caste, ethnicity, gender or location. The programme realized that many of the poorest households were excluded from membership in the community forest user groups, decision-making, access to the forest or access to other financial or infrastructure benefits. The strategy was introduced at every level, from user sub-groups to the ministry (ref. 4a).

## Terai (BISEP-ST) (2005–2010)

The Biodiversity Sector Programme for Siwaliks and Terai (BISEP-ST) was a programme of Nepal's Ministry of Forest and Soil Conservation, and was funded by the Government of the Netherlands and managed by SNV Nepal. The goal of the five-year programme (2005–2010) was to work toward a self-sustaining forestry sector in Terai, inner Terai and Siwaliks for biodiversity and equitable economic development. The main approaches used to achieve this goal were to strengthen the forest sector institutions at different levels, to promote decentralized and community participation, and to ensure the equitable distribution and reinvestment of forest resources. The programme was implemented in the eight districts of the Terai and inner Terai of the country, where it benefited 440,000 poor farmers through the management of 5,000 hectares of forest land in eight districts (ref. 21a).

Initially BISEP-ST focused on sustainable forest management as a means for achieving biodiversity conservation and economic development objectives. Learning from its initial interventions, emphasis was given to communities and their representative institutions. At local level, the programme focused on strengthening local forest user groups (of different kinds), enabling them to deliver livelihoods benefits for their poorer and socially excluded members. At district level, the programme emphasized strengthening accountable and representative institutions, especially the District Forest Coordination Committee, which was put at the heart of the programme to support devolution (administrative and fiscal) of the forest sector.

BISEP-ST had an explicit emphasis on addressing issues of poverty and social inclusion within the forest sector – particularly among the poorest members of forest user groups and other forest-dependent people. The programme worked to support the implementation of those forest sector programmes that could best demonstrate an ability to contribute to the overall development objective. In addition, BISEP-ST focused on conflict sensitivity peace building, aiming to strengthen those aspects that can contribute to addressing the causes of Nepal's conflict (and hence, an increased emphasis on addressing poverty and social exclusion, and promoting good forest governance).

The programme continued to be led and managed from within the Ministry of Forest and Soil Conservation, since it incorporates activities and actors from across several departments within this ministry. However, during the extended period of the programme there was greater emphasis on close coordination with other relevant ministries, notably the Ministry of Local Development and the Ministry of Finance, thus reflecting decentralized governance (ref. 3a).

## The Multi Stakeholder Forestry Programme (2012–2022)

> "We are already replicating the leasehold forestry programme [in eight districts]. It is part of the Department of Forestry's Multi Stakeholder Forestry Programme, currently in its first phase. We are developing the programme focusing on people living below the poverty line and discussing how the forestry sector could help their livelihoods. Recently the Ministry of Forest developed a vision – Forestry for Prosperity – which is about how the forestry sector can bring this about. Based on discussions with the Ministry of Finance, we can continue the programme with public financing. This is our commitment.

The Multi Stakeholder Forestry Programme works in 23 districts and eight of them are districts involved in the Leasehold Forestry Programme. The team from the forestry programme intends to continue working with these common districts. They are also going to extend to more districts, from 23 to 63 after 2014, when the second phase begins and for which funding is already assured from DFID, and the governments of Finland and Switzerland. MSFP will be supporting leasehold forestry as well other forest management modalities. This is how leasehold forestry is being replicated in other districts and to other programmes. The new programme will also seek ways to integrate community and leasehold forestry. We have different policies to run community and leasehold forestry, so while they could be integrated they cannot be merged. It also depends on the willingness of the users as well. For example, if ten members of LFUGs see their economic situation rise from one of poverty to a higher level after ten years, they may want to have their lands brought into the community lands, and there might be options. If they rise from poverty level to a higher level, they may want to have their lands brought into the community lands."

*Mr. Biswa Nath Oli, Director-General of the Department of Forests under the Ministry of Forests*

The Multi Stakeholder Forestry Programme (MSFP) aims to improve the livelihoods and resilience of poor and disadvantaged people. It will also develop the contribution of Nepal's forestry sector to inclusive economic growth, poverty reduction, and addressing climate change.

The MSFP is the product of a multi-stakeholder design process undertaken in Nepal's forestry sector. As such, it the first of its kind and builds on the achievements of over 20 years of forestry work of the government. The programme is funded by the Government of Finland, the Swiss Agency for Development and Cooperation (SDC) and the UK Department for International Development (DFID).

The MSFP's main beneficiaries are rural communities, especially poor and disadvantaged households, and those most vulnerable to climate change. The programme seeks to bring an estimated 1.7 million people out of poverty by working with existing and new forestry groups of various kinds and creating an additional 80,000 jobs. Four outcomes are anticipated by the end of the programme:

- Government and non-state actors (civil society, NGOs, communities and the private sector) jointly and effectively implement inclusive forest sector strategies, policies and plans.
- The private sector (farmers, entrepreneurs and financial institutions) increase investment and jobs in the forestry sector.
- Poor rural communities benefit from local forest management and other investments.
- Forest and trees are sustainably managed and climate-resilient, and are monitored by government, communities and the private sector.

One new area of endeavour is the role of forestry in helping communities to adapt to climate change and mitigate its impacts, and the piloting and development of payments for environmental services (such as forest carbon financing) that bring additional benefits to local communities and the nation.

In January 2012, the Government of Nepal and three donors signed a joint funding agreement to implement the MSFP for the first four years. A second six-year phase is envisaged.

The programme is being implemented by a wide range of stakeholders– the government, NGOs, civil society and the private sector – to strengthen national, regional and local institutional arrangements that can deliver effective forest sector development.

## Results expected

The following results are expected over the next ten years:

- Contribute to lifting 1.7 million poor and disadvantaged people out of income poverty.
- Reduce the climate vulnerability of 560,000 households.
- Increase the contribution of the forest sector to Nepal's GDP from 9 per cent to 10.4 per cent.
- Double the contribution of forest-based activities from 3 per cent to 6 per cent of household income.
- Halve the deforestation rate in programme districts from 1.7 per cent to 0.8%.
- Improve forest sector governance and support the establishment of a multi-stakeholder "national entity".
- Increase private sector investment in the forestry sector and create an additional 80,000 jobs.
- Increase the area of forest managed by local forestry groups by 100,000 hectares.
- Double the area with improved forest conditions. (ref. 10)

**A strategic partnership with the International Centre for Integrated Mountain Development (ICIMOD)**

IFAD and ICIMOD have had a strategic partnership for nearly a decade. ICIMOD's core strengths are in issues relating to livelihoods and climate change and adaptation. The ongoing grant is about that, and about the capacity of communities to deal with change – not only climate change but also socio-economic change. Our long-term partnership has been with WUPAP and LFLP.

DABUR Nepal, a private agency from India, was also brought in. They deal with ayurvedic medicines and the like. When you talk about higher-altitude areas, the options are limited, agriculture is limited, the area is fragile. But medicinal plants and NFTPs can be developed. The arrangement with DABUR was to select certain plants that have a market demand and to test them. The first phase was tested for six months. We had four plants that were suitable and two were doing well. We set up greenhouse with nurseries, involving the communities. The next phase was upscaling or outscaling. Outscaling is replication of what you are doing. Upscaling it is an approach which the next level of decision-making takes up. The scale has increased. The next-stage species were selected, and along with them capacity building, expansion, and linking to markets. One of the lessons we have learned is the importance of the private sector. What flexibility is there in terms of setting prices and marketing products? When you enter into a contract you fix the price, but down the line the market price could be much higher. The grower will naturally want the higher price, and a breach of contract can occur. This comes about because of the lack of flexibility. Another lesson learned is the importance of bringing in semi-processing. Just dealing with raw products, producers get a miniscule percentage. Small-scale commercialization needs to be taken up in future projects.

One area of concern when you are working on slopes is management of soil – water management, soil erosion, etc. We addressed these aspects with an NGO partner and LFLP, since the latter is working with degraded plots. Again, we had some trials and demonstration in the field with farmers to explore how you can manage the soil, how you can enhance fertility, how you can control soil erosion in sloping areas. We introduced contour hedgerow, which is a simple technology where hedges are planted along the slopes. Instead of physically creating walls, there are varieties of nitrogen-fixing plants and those that the community uses. Thus, they can use forage species as hedges and fodder for their livestock, without having to travel far to collect it. This cuts down on drudgery of women and helps in erosion control. And this way they were managing the plots that were there. We applied this technology together with LFLP project team members, the district forest offices, and rangers, who were trained so that they could disseminate the technology in different districts. We also conducted an assessment to look specifically at the community perceptions of change, and impacts that the communities were experiencing. What livelihood activities were being impacted? Could they cope with that? Extensive surveys were conducted in five districts, six villages in each district. That assessment gave us a very good idea of what needs to be improved.

Of relevance to LFLP is the need for landscape management of resources, particularly in regard to water management. Water stress is the main issue. In addition, most of the communities who had livestock were reducing herd size, and shifting from bigger to smaller animals. This was not just for reasons of climate change, market demand, etc., but to enable them to obtain immediate cash and high returns during certain times of the year. We started some veterinary health services through the communities to ensure the health of their livestock.

A specific intervention that we did was in regard to the potential of carbon trading. Everyone is talking about carbon trade, REDD plus, and Nepal has taken a lot of steps. Our point was that when you talk about leasehold forestry, the plots are quite degraded and they are being leased to poor people, who need an immediate return. However, you also need to think long term and to build capacities of groups to measure carbon. If you can show a steady inventory, then the communities can demonstrate the carbon sequestration of a time period. Another argument after we have the results is that the principle of carbon sequestration depends on rates of photosynthesis. With leasehold forestry, the forests are young, so as they regenerate, the rates of photosynthesis and sequestration are going to be higher. We need to prove this. We need the build capacity of communities to measure. Then we need to show that over a time period the sequestration is better. Moreover, when leasehold forestry begins, there isn't tree cover. There are herbs and shrubs, which are fast-growing annuals. The sequestration in annuals should be higher, which is another added advantage. Since they are annual, the turnover of the biomass through decomposition will be higher and faster. Therefore the soil carbon will be higher. These are all hypotheses at the moment. We are testing them. In the meantime we are building the capacity of project teams and staff to conduct the analysis and mapping, and of communities to measure.

*Dr. Dhrupad Choudhury, Programme Manager, Adaptation to Change & Initiative Coordinator, ICIMOD*

## Other uses of the leasehold concept

Leasehold farming is an innovative concept that has already had positive successes in Nepal. NGOs have been implementing projects for private leasing of crop land and for farming in riverbeds.

### *Leasing of private land*

The UK-based NGO Practical Action implements livelihood support projects in Nepal. In 2006 it initiated the project Improving Livelihood Security of Socially-excluded Communities in Nepal, which was completed in 2010. The objective was to increase the income of the land-insecure, vulnerable and socially excluded households by diversification of livelihood options in six conflict-affected districts in Western Nepal.

The target group was comprised of low-caste and marginalized communities such as *dalits*, minority ethnic nationalities and households headed by women that possess less than 500 $m^2$ of land. The project was implemented in partnership with the Local Initiative for Biodiversity Research and Development and the Dalit Welfare Organisation in Nawalparasi, Rupandehi, Banke, Surkhet, Kailali and Doti districts. It also worked closely with the respective District Development Committees, Village Development Committees and line agencies.

The leasehold farming component supported households to form small groups of about ten members and then helped them to negotiate private lease contracts with landowners, initially on a one-year basis and subsequently for longer terms. Similar to the Leasehold Forestry and Livestock Programme, it offered group members training and inputs for a productive use of the leased lands.

In 2007, one year after the start of the project, Practical Action reported that 680 farmers from 72 leasehold farmers groups were currently leasing 45 hectares of arable land for a four-year period. Most of the land was being used for vegetables and cash crop production, and two to three harvests were being obtained. Groups invested part of their profits in leasing larger areas in following seasons.

The project provided support to leasehold farmer groups for four years, during which the external funding of rent and inputs declined from 100 per cent in the first year to 25 per cent in the last year, with the groups mobilizing the remaining 75 per cent. In addition, irrigation was developed in the leased lands through treadle pumps or shallow tube wells with electric or diesel pumps. Training in vegetable growing was provided in the beginning and refresher courses were given as required.

After the project intervention, income of the beneficiaries had increased by 49.2 per cent. This resulted from the marketing of vegetables produced in the leased land. It was estimated that households gained a net income of NPRs 37,000 (USD 370) per year after deducting all production costs.

Most household income before the project was earned through farm labour. After the project intervention, the contribution of crop production to household income increased from 5.4 per cent to 29.9 per cent. With the acquired knowledge and skills in improved vegetable farming, farmers also have the option of growing vegetables on their own land or on leased land. This indicates that the farmers are moving towards self-employment.

Some 58.3 per cent of the households had food security for less than three months before the implementation of the project. After the project, it had fallen to 6.7 per cent.

### *Riverbed farming*

Many poor people in Nepal have been moving from the hills to the Terai. The reasons for migration are many: the available land for cultivation is getting smaller as farms are divided among children; environmental impacts on the land; and growing poverty. At the same time vast tracks of riverbeds are dry and fallow from October to May. It is common land under the responsibility of village authorities. The land is fertile, yet it is currently not used.

The NGO Helvetas Swiss Intercooperation operated a small pilot initiative in riverbed farming in 2007 and 2008, including around 1,000 households in Far Western Districts of the Terai and financially supported by DFID. After winning the Development Marketplace 2008 of the World Bank with a proposal on Riverbed Farming, Helvetas was able to expand its activities during the 2008–09 cultivation season and supported 2,200 landless and land-poor households in Kailali and Kanchanpur District in their effort to take up riverbed farming.

The riverbed farming project aims to develop a system through which resident landless and land-poor people gain access to riverbeds during the dry season. They are then able to cultivate horticultural products, link up with local and regional markets and generate income. The system developed includes:

- allocating land and selecting appropriate sites in dry riverbeds;
- identifying primary stakeholders constituting groups of women and men farmers;
- preparing and signing leasehold agreements with the respective Village Development Committees, which assure them the right to cultivate a specific area of dry riverbed for the coming three years;
- providing agricultural extension through about 20 project-trained local resource persons (whose services are registered as private business and linked to the Micro Enterprise Development Fund in order to ensure long-term sustainability; and
- marketing the produce through links with traders and wholesalers.

A National Riverbed Farming Strategy plan is being developed together with the Ministry for Local Development and the Ministry of Agriculture and Cooperatives, and further elaborated by the Riverbed Farming Alliance, which was established in 2011 and includes as members: the Forum for Rural Welfare and Agricultural Reform for Development; the Deutsche Gesellschaft fuer Internationale Zusammenarbeit; HELVETAS Swiss Intercooperation; Mercy Corps; the Poverty Alleviation Fund; the Ministry of Industry; and the Micro-Enterprise Development Programme of the UNDP. Apart from work on the strategy, the Alliance also created a web page, organized field visits and workshops, developed Riverbed Farming Guidelines, undertook riverbed mapping in four pilot districts and undertook participatory action research on micro and small irrigation technology.

Riverbed farming has successfully increased household income and improved the food security of landless and land-poor households in the Western Terai of Nepal. A total of 122 landless and land-poor farmer groups, including up to 3,050 households, were engaged in riverbed farming. In the last season, average production and income per kattha of land was 510 kg (15,100 kg/ha) and NPRs 6069 (NPRs 179,500/ha, USD 1,795), respectively. It is estimated that on average a household earned a net income of NPRs 24,276 (USD 243) from riverbed farming, which constituted 35–40 per cent of the household's total income. Riverbed farming produce is also bartered with food grains, which fulfilled four months of additional food grain requirements of the households (ref. 20 and 4).

# Chapter 6

## When the programme leaves

*In Jhirubas, Palpa district, the well-tended dirt road, the village's only artery, seems to be suspended in air. On either side, the land slopes sharply downward, with only the thatched roofs of the houses at waist level. Wherever the eye turns, one sees only sky, forest, cultivated fields, and tiny etchings in the slopes – roads. Yet along this main road in Jhirubas, one perplexing anomaly stands out: Perched on a wooden pole is a solar panel the size of a small flipchart. This stark and singular contrast to the surroundings provokes many questions: How did it get here? What is it for? And, most importantly, how will it be maintained? For one day soon, a complex dismantling will begin to take place. Certain people, tools and pieces of equipment will be removed from the scene and used elsewhere, taking with them not only their pragmatic utility but certain flows of knowledge, innovation and energy. The operation will be an extremely delicate one. With so many interlocking pieces, it's nearly impossible to know which one/s, if any, could cause parts of the structure to come tumbling down once they are excised. Not unlike a house of cards, except that these are real houses, with real people living in them, who have experienced the depths of poverty and have started to gain the skills to lift themselves out of it. No, this is no house of cards, and the famous "exit strategy" cannot be a roll of the dice.*

The most important question is: what happens when the programme leaves? Will leasehold forestry be able to forge ahead on its own momentum and be further embraced and supported by the government and the poor people who are deriving benefits from these forest areas? Will it gradually disintegrate without the external support that has been given over the past two decades?

This chapter explores the issues of sustainability and of the integration of community and leasehold forestry. What needs to happen and how? Who needs to make it happen and how? And, most importantly, can it happen? The sections presented in this chapter are adaptations of "thought pieces" that were written expressly for this chapter by those who have been intimately involved in leasehold forestry since the concept first took root. The chapter ends with a summary of the Kathmandu Declaration June 2014 arising from LFLP closing workshop: "the Regional Workshop on Pro-Poor Leasehold Forestry" which brought them and over 170 other practitioners, leasehold farmers, decision makers, academics, civil society – with over 20 years of collective experience – to discuss the future of leasehold forestry. Let's hear what they have to say.

## Looking ahead[25]

Since its introduction more than two decades ago, leasehold forestry in Nepal has crossed a number of milestones. In terms of longevity, it has already crossed over as a priority programme. In terms of breadth, it has been adopted in over 40 districts with around 7,000 groups benefiting over 40,000 poor families. In terms of acceptance, the concept has been adopted by other projects and programmes such as WUPAP, BISEP-ST and LFP, and has also been mainstreamed into community forestry in the form of land allocations, leasehold forestry for public and institutional lands, etc. The concept has also bolstered environmental conservation, increasing biodiversity and forest production.

Progress and achievements over the two decades have been quite remarkable. Nevertheless, there are two particular issues that need to be taken into highest consideration for the future of leasehold forestry: how can leasehold forestry be continued and expanded sustainably; and how can it be mainstreamed into other community-

[25] By Bala Ram Adhikari, Programme Coordinator, Leasehold Forestry and Livestock Programme.

based forest management tenure systems in Nepal, such as community forestry?

## How can leasehold forestry continue and expand in a sustainable way?

The foundations of leasehold forestry are the leasehold forest user groups (LFUGs). Many of these groups still live below the poverty line and will need additional and longer support if they are to mature into sustainable groups. A recent study has shown that 21 per cent of the groups are very active, 57 per cent are medium and 22 per cent are passive. This fact provides an exciting scenario because the results drawn include the groups formed in first phase (1992–2002). Further support to the active groups and some to medium groups, to enable them develop enterprises or other commercial activities, could be one of promising areas for future interventions. The more passive groups could be revived with little support.

One important feature that the programme envisioned to promote the sustainability of LFUGs is the concept of the Livelihood Improvement Plan, which is the principal guiding document for each group towards its sustainability. So far in 2013, Livelihood Improvement Plans have been prepared for only about 3,000 new groups. The remaining groups, and most first-phase groups, need to review their Operational Plans with a view toward integrating the Livelihood Improvement Plans into them. This would bring a number of advantages: (i) revision will ensure that the socio-economic and bio-physical requirements of LFUGs in a changing context are being addressed; and (ii) opportunities will open up to include "climate smart" activities to address climate change adaptation and mitigation. However, LFUGs will need support by other development partners, in partnership with the government, in order to successfully create and implement their plans.

The groups will also need support to help them establish themselves as sound institutions with good governance. Governance is especially important as the groups continue generating financial resources either in the form of group "saving and credit funds" or in the form of cooperatives as a networking function for these funds. Only the highest level of governance will ensure that each of the members benefit. Thus, these groups should be guided, by donors and the government alike, for next few years to make them self-reliant.

Leasehold forestry has become a tested model of success for reducing poverty and improving the environment. The Government of Nepal has recognized this success by awarding LFUGs with Environmental Conservation Awards, Mountain Development Awards and other explicit forms of recognition. Therefore, this programme should be scaled up in the remaining districts of Nepal where there is potential for practising leasehold forestry in one form or another. The Ministry of Forests and Soil Conservation (MoFSC), and especially the Department of Forests (DoF), should lead the national leasehold forestry programme. A strong institutional set-up at the higher levels of the MoFSC and DoF needs to be in place in order to translate the key lessons learned and to further incorporate leasehold forestry into the existing Forest Acts and Regulations.

"There are two departments directly involved in the programme and legally responsible for conducting activities – the Department of Forests and the Department of Livestock Services. Apart from them, there are many other organizations involved that are important in making the programme a success. However, trying to hold coordination meetings is always a difficult job. Moreover, so many people involved in the programme need capacity development. The District Education Offices have several programmes for formal and informal education. Education is the only thing that helps. Once people are capacitated, they can continue with income-generating activities. Therefore, the education stream should also be involved in our programme implementation approach.

"Many people are illiterate and have little education. They need more capacity development activities and *regular* support – for example, in holding monthly meetings, keeping their records, completing forms, and so on. There are thousands of groups, and each group has about ten members. Add it up and it is a lot of people. We have about one social mobilizer for every 170 groups. Moreover, many forestry and livestock staff have little knowledge of rural finance or finance management. It's an enormous task.

In addition, we have tried to establish a Memorandum of Understanding with financial institutions, but again there are some differences between their modalities and the aspirations of the beneficiaries, especially in terms of interest rates, which can be as high as 25 per cent. These are big challenges for sustainability."

*Mr. Biswa Nath Oli, Director-General of the Department of Forests under the Ministry of Forests*

FAO's piloting of leasehold forestry in four districts (through its Technical Assistance) has shown the best results with an integrated approach at the landscape level, which needs to be replicated in other areas of the country. A package of support from resource generation to marketing of products can make leasehold groups break out of the poverty line.

## How can community and leasehold forestry be better integrated?

Currently, 18,000 community forest user group (CFUGs) include poor and vulnerable people to some degree as members, and part of their land holdings include fallow, unproductive, under-utilized lands that have huge potential for land allocation. Integrating the two approaches has already been adopted by other projects (for example DFID/LFP and SNV/BISEP-ST). In addition, many community forestry guidelines envision that 35 per cent of a CFUG's annual income will be used toward pro-poor and social inclusion activities. More recently, FAO has been piloting integrating community forestry and leasehold forestry in two districts. The lessons learned could be extremely valuable to scale up this integration effort in many CFUGs.

At the same time, it must be remembered that the Department of Forests manages community forestry and leasehold forestry as separate programmes. Despite their many similarities, they are not yet well integrated, resulting in several practical problems and complexities at implementation level. The Forest Act 1993 has given priority to community forestry over leasehold forestry. Therefore consensus of the community is required before handing over the leasehold forestry – and more often than not, the community is dominated by several elites. Moreover, community forestry started in Nepal in the late 1970s. Leasehold forestry started only in the early 1990s, as discussed in Chapter 1. Most of the accessible forests have already been handed over as community forestry. Therefore obtaining unallocated government land for leasehold forestry is limited, and studies have shown that leasehold forestry beyond 3 km of a community is not practicable. (This is the reason that Leasehold Forestry and Livestock Programme was not able to meet its area handover target even though it exceeded its group formation target).

However, there is scope for increasing the leasehold areas:

- The threshold up to which leasehold forestry could be handed over is up to 20 per cent of crown cover. The Kathmandu Declaration 2007 has appealed to the government to increase this threshold, with a focus on poor people not on degraded forest land. If this focus from degraded forest is shifted to poverty pocket areas, then up to 50 per cent of crown cover forest could be handed over.
- The leasehold area could also be increased if community forestry and leasehold forestry should ever be legally on an equal footing, or at least seen more strongly as complementary approaches.
- The districts where leasehold forestry is being implemented could be extended beyond the current districts. As a programme, the concept can be expanded to any districts provided there is

sufficient funding to implement and monitor activities, and to provide training and other inputs. In addition, models would have to be developed that take into consideration specific topographical aspects of the area, such as elevation and rainfall patterns.

- Other potential leasehold sites can be possible inside community forestry lands – as a number of leasehold plots did evolve under LFLP. It is roughly estimated that about 5 per cent of the whole community forest land area have potential for pro-poor leasehold forestry, in terms of degraded areas. Such an area would be equivalent to the existing leasehold areas (about 50,000 ha). This would of course be subject to considerable negotiation with community forest user groups, and much may not be suitable, but some of the degraded areas would be perhaps willingly allocated (see more below on integrating models).

Separate management of community and leasehold forestry has also increased the cost and time required for the leasehold forestry programme to be implemented. If DFO staff are oriented on both concepts, it saves time and the cost of the training and also resolves the staffing problem. It must be remembered that when CFUGs are being formed, they conduct a wealth ranking, which resolves the issue of identifying the poorest. In addition, almost 80 per cent of LFUG members overlap with CFUG members. LFUGs often do not have enough funds to disburse loans to their members but have to depend on external sources for the credit fund. With community forestry's provision of investing 35 per cent of its income in pro-poor activities, implementing the two approaches jointly could enable the community forestry fund to support the poor for 35 per cent.

## Models for integrating community and leasehold forestry[26]

Two refined pilot models for integrating community and leasehold forestry were launched in 2012. One model includes leasehold forests inside community forests (Bandeu, Sindhupalchok), and the other model uses land separated from community forestry and handed over as leasehold forest (Tandrang Taxar, Lamjung). The results of these models should pave the way towards integrating both approaches on a larger scale. The processes followed in each model are outlined below, along with the key issues, lessons learned and recommendations.

### *Model I: Leasehold forestry inside community forest*

- Formed two sub-groups involving all 64 households of two settlements having similar well-being status. The size of the sub-group was much larger than the normal LFUG to make collaboration with the Poverty Alleviation Fund possible.
- Carried out forest boundary survey of the allocated land in the presence of CFUG representatives and prepared a map. Area allocated for the poor households was 28.3 hectares.
- Prepared household-level Livelihood Improvement Plan of both sub-groups.
- Transferred tenure right from CFUG to sub-group of poor households through an agreement between CFUG and sub-group.
- Prepared land development plan and Livelihood Improvement Plan for each sub-group.
- Organized several coordination meetings with development partners to pool resources for Livelihood Improvement Plan implementation.

---

[26] By Govinda Prasad Kafley, Team Leader, FAO Technical Assistance project for Leasehold Forestry and Livestock Programme.

- Organized joint field visit of district line agencies and political leaders to the pilot site.
- Conducted several trainings such as vegetable cultivation, livestock rearing, sericulture and mulberry cultivation, saving and credit, record keeping, goat rearing and Start and Improve Your Business (SIYB) for sub-group members.
- Organized land development coaching for each sub-group. Both men and women from each household were involved in land development coaching.
- Provided planting materials (fodder – 22,000 seedlings; fruit seedlings – 2,500; multi-purpose tree species – 8,000; and forage seeds).
- Provided Livelihood Improvement Plan implementation support to each sub-group.
- Supported water pump for lift irrigation.
- Initiated monthly saving (NPRs 50 per household) in each group.

***Model II: Community forest handed over as leasehold forests returning part of its area***

- Decided to provide part of the community forest area for leasehold forestry implementation by the general assembly of CFUG.
- Delineated the forest area for leasehold forests in the presence of CFUG representatives. The area of the forests thus delineated was 15.5 hectares.
- Revised the operational plan of the community forests. A total of 15.5 hectares of the forest area was deducted from the community forests. A new provision was added in the operational plan of community forest to allocate 35 per cent of its income for pro-poor activities.
- Carried out well-being ranking to identify the poor households within CFUG membership.

- Submitted the operational plan of community forest to DFO for approval after the endorsement of the CFUG general assembly.
- Initiated the process of LFUG formation.
- Formed three LFUGs involving 20 poor households (two groups with seven households and one group with six households) identified from well-being ranking.
- Prepared operational plan of all three leasehold forests with Livelihood Improvement Plan incorporated into it.
- Submitted the operational plans to the DFO for approval after the endorsement of the LFUG.
- Issued a lease certificate to each group after thorough review and approval of the operational plan by the DFO.
- Organized land development coaching in each LFUG. A total of 36 LFUG members were trained on sloping agriculture land technology.
- Provided planting materials (forage seeds – 11 kg; napier – 2,000 sets; broom grass – 5,000 rhizomes; fodder tree species – 900; fruit seedlings – 120) to each LFUG.
- Organized forest management coaching to each LFUG. A set of tools for forest management was also provided.
- Distributed a pair of goats to each household and one breeding buck to each group.
- Conducted goat rearing and shed improvement training with 20 LFUG members.
- Organized a one-day animal health campaign and treated 209 goats for parasites.
- Organized coordination meetings to pool resources for Livelihood Improvement Plan implementation (resources pooled to construct conservation ponds and toilets).
- Initiated monthly saving (NPRs 20 per household) in each group.

### *Key issues*

1. Difficulty in getting community consensus in favour of poor households.
2. CFUGs reluctant to provide productive part of the community forests to poor households.
3. Long-term security of tenure in leasehold forestry inside community forest is still difficult.
4. Working with poor and disadvantaged groups is time-consuming and requires rigorous work. The present strength of the DFO is not enough to support community forestry and leasehold forestry.

### *Learning*

1. Individual allotment (informal arrangement) of lease plots develops strong sense of ownership and responsibility, resulting in better management of the lease land.
2. Integrating community forestry and leasehold forestry manages the problem of land availability for leasehold forestry.
3. Leasehold forests implemented in the poverty pocket area (settlement basis) is less likely to cause conflict.
4. Integrating forage development activities in forestry increases women's participation in forestry development activities.

### *Recommendations*

1. Leasehold forestry should be implemented in the poverty pocket area (settlement basis) to resolve the issues of community consensus.
2. An obligatory legal provision should be in place allocating certain parts of the community forest area on long-term lease for the benefit of poor households.

The leasehold users and community forest users almost overlap in most cases. Generally, as mentioned earlier, around 80 per cent of the leasehold users are community forest users as well. The leasehold users have specific needs regarding their livelihood strategy. The users satisfy their needs from leasehold forests. Community forests, which are larger and have a large number of users with diversified needs, cannot address the specific needs of individual households. Thus leasehold forestry caters to the specific needs, and community forestry caters to the common needs.

The leasehold forests generally are developed in barren lands around the communities, which almost always ban grazing. This acts as a buffer to the community forests and checks deterioration of forests. Similarly, with intensive care and fewer fire hazards in leasehold forests, often forest fires are prevented from spreading into the community forests from the villages.

The concept of community forestry forming a sub-group of users who fall below the poverty line has been introduced successfully in many groups, and there is great potential to expand this concept. Sometimes very enthusiastic community forest groups have gone to the extent of reverting back to the government a part of the land already that had already been handed over to them, so that it can be later handed over to their poor households as leasehold forestry. This ensures long-term tenure and increases harmony among the community.

On a more anecdotal note, in Palpa district, at first the people wanted to chase us [the government FAO TA team] away, and now they welcome us. The change is there. When you go to Palpa now, it is almost impossible to envision what it was like before, three years ago. The achievement is something. We recently brought the secretary of planning for a visit. Afterward, he said, "If you had given me documents saying that this is what you had done in three years, I wouldn't have believed it." Yes, we need to go forward, to move ahead.
*Govinda Kafley LFLP TA Team Leader*

## 20 years of leasehold forestry: putting experience to good use[27]

The leasehold forestry approach is now well established in Nepal and considered one of the most innovative and widely recognized programmes for combating poverty and rehabilitating degraded forests in the Hills. Significant achievements have been made in natural resource management, food security, microfinance and other components of rural livelihoods such as community infrastructure and institutional development. Furthermore, the programme has been able to bring policy implications and spill-over impacts in favour of poor rural people.

Despite these positive outcomes, a number of institutional, technical and socio-economic issues need to be addressed to ensure that the leasehold forestry concept is sustainable and can be scaled up. The issues frequently cited by researchers, resource managers, sociologists and donor agencies include: a "blanket" approach that is mostly donor-driven; competition with community forestry; unsecured tenure rights and conflicts with customary land tenure; exclusion of the poorest households; lack of support after groups have been formed as well as for livelihood diversification and value chain development; poor coordination and governance on the part of government and LFUGs; and inadequate monitoring and knowledge management systems. In my personal and professional experience with leasehold forestry of more than 20 years, I would like to add several issues that I believe are fundamental: deteriorating professionalism among the foresters and heavy reliance on donor agencies; limited and unorganized number of champions or change agents; traditional approaches to planning rather that planning based on theories of change; lack of risk management and inadequate attention to pathways for scaling up; and a highly fragile political environment where vested interests are at play.

---

[27] By Nav Raj Baral, National Expert, FAO, Nepal.

### *Lessons learned*

A number of important lessons have been learned in the leasehold forestry experience to date. They are briefly described below.

***Policy considerations.*** Given the well-defined property rights and simple local knowledge-based land-use technology, degraded forest cannot only be restored in a short period of time, but sustained production for consumption and commercial uses can also be achieved. At the same time, livelihood improvement and environmental conservation are complex and dynamic processes. One-time policy formulation is not enough. Policies need to be continually revisited and amended to respond to socio-economic, political and environmental changes. A well-established knowledge management system enriched with self-monitoring is a must to influence policy makers, politicians, donors and civil society, and to make policy conducive to leasehold forestry.

***Integrating forestry with other livelihood options.*** Integrating forestry with a livestock-rearing programme is crucial to generate early income and lay the foundation for livelihood improvement. However, it is not a panacea for sustainability. In an agrarian society like Nepal, breaking poverty and conserving the environment sustainably requires a multi-disciplinary and integrated approach that goes beyond the current level of involvement of a few development sectors (i.e. forestry and livestock).

***Establishing strong pro-poor-institutions.*** Establishing strong pro-poor institutions is essential to take the ownership of interventions and sustain their outcomes. Changes need to take place in working behaviours and bureaucratic norms on the part of government and development partners in order for internal resources to be mobilized to carry initiatives forward in the absence of projects or external support. Establishing market linkages for LFUGs and setting up appropriate value chains are two of the best options to maximize benefits to LFUGs and promote sustainability.

***Developing synergy with community forestry.*** Taking into account the distribution of forests, leasehold forestry cannot – and should not – compete with community forestry. Experiences from community forestry indicate that it is unable to efficiently address the livelihood needs of poor and marginalized communities. A synergy between community forestry and leasehold forestry is urgently needed in which these two modes of forestry complement the interests of one another and at same time contribute to the national goal of poverty alleviation and sustainable management of forests.

***"Reinventing" forestry agencies.*** Re-orientation and role change are painful, as maintenance of the status quo is always the preferred tendency. The identity of forestry professionals and technicians, and the public perception of them, is important. Even after 30 years of community-based forestry, civil society groups and donor agencies are distinctly divided. Visionary leadership, committed political support, and an ability to win followers and influence detractors are critical in bringing about positive change in Nepal.

## *The way forward*

Nepal has been in a political vacuum for the last ten years. The Constituent Assembly, elected in 2008, failed to write a constitution. Political parties could not hold new elections. Therefore, the Chief Judge of the Supreme Court was sworn in, in early 2013, as head of an interim government made up of a cabinet of retired and senior-most bureaucrats. (The interim government has been able to hold elections successfully in November 2013. A coalition government between the Congress party and the United Marxist-Leninist party is in force.)

However, other components or drivers of sustainability and scaling up are inspiring. The policy environment seems to be conducive to community-based forestry and pro-poor forestry. The national agriculture policy, known as the Agriculture Development Strategy (ADS), was finalized in 2013 and will lead the sector for the next 20 years. The forestry sub-component of the Agriculture

Development Strategy has strongly recommended a competitive and agriculture-friendly forest policy and stresses the expansion and/or mainstreaming of the leasehold forestry concept into government-managed forests, community forests, watershed management and protected forest in a more integrated way. The time frame for the 20-year forest sector policy (known as the Master Plan for the Forestry Sector) has expired, and a new forest policy is being formulated. This is the right time to address the second-generation (social equality and governance) and third-generation (global environment) issues.

In addition, despite the political turmoil, the majority of donor agencies working for forestry in Nepal have joined together under a highly ambitious and debated programme known as the Multi Stakeholder Forestry Programme (MSFP). The ten-year programme, signed in 2012, aims to bring an estimated 1.7 million people out of poverty by working with existing and new forestry groups of various kinds and creating an additional 80,000 jobs. Poverty reduction and tackling climate change is its motto.

Taking into account the lessons learned and the socio-political scenario of the country, the time has come to move forward with more context-specific strategies and actions that will scale up leasehold forestry and promote its sustainability and the development of a pro-poor forest policy. These strategies and actions fall under two broad areas:

### *A. Consolidate the achievement of leasehold forestry and strengthen the institutional and technical capacity of the forestry institutions*

#### A.1. Develop a strategic plan for consolidating LFUGs and for scaling up

- Develop a consolidation or scale-up plan for each category of LFUG (active, moderately active) taking into account the pathways of change, site specificities and livelihood assets.

- Revisit the Livelihood Improvement Plan process in line with the theory/pathways of change and make it a road map for change.
- Help develop a district forestry sector planning framework that mainstreams and integrates leasehold forestry into the overall district development planning framework.

A.2. Develop a site-specific, competitive, productive and local knowledge-based land use model

- Develop, pilot and scale up various multiple-use site-specific forest land-use models for leasehold groups, integrating them with farming and value chain development systems across the agro-ecological or landscape zones of the country.

A.3. Develop and promote integrated value chain development

- Analyse mountain specificities of leasehold forestry areas and work with LFUGs to choose appropriate value chains based on accessibility and market linkages.
- Conduct value chain analysis and prioritize strategic products (including unique and niche products) for each site/area, determining whether they are low-volume high-value, or high-volume low-value.
- Help organize LFUGs into groups, networks or cooperatives and support them to develop and implement a value chain development strategy.
- Work to create an enabling business and policy environment in partnership with the private sector.
- Develop participatory self-monitoring and evaluation system with well-defined indicators and document/disseminate lesson learned.

A.4. Create operational guidelines on developing forest and community-based forest/agriculture enterprises

- Work to reclassify forest products as agricultural products in order to exempt them from the permit process for harvesting and business, and from royalties and taxes.
- Work with the Department of Forests and/or the District Forest Office to develop linkages to apex financial institutions of the government and other commercial financial institutions and resources.
- Develop pro-poor or community-based public–private partnerships for rural credit and value chain development.

A.5. Develop LFUGs into strong pro-poor institutions

- Federate LFUGs into networks (intergroup), and networks into cooperatives and work with them to formalize themselves as business entities and in order for them to be able to lobby and influence policy makers and government line agencies.
- Revisit the overall institutional performance of LFUGs in terms of their institutional and technical capacity and develop plans to further strengthen each category (active, moderately active and passive).
- Provide necessary support to the LFUGs and their networks for shared learning.

A.6. "Re-invent" forestry institutions

- Assess the effectiveness of the various institutions established at the department, district and project levels and restructure them to make them more leasehold forestry-friendly.
- Conduct regular awareness-raising/reorientation training packages for different types of stakeholders (local, district, region, department or policy level) to have a better understanding and shared vision about the context, international commitments

and obligations, and policy implication regarding leasehold forestry.

- Review and refine the existing forestry sector capacity enhancement training to make them context-specific, pro-poor and results-oriented.
- Work to equip forestry organizations with necessary physical facilities and incentive packages to enable them to proactively respond to the development of leasehold forestry as well as to become "learning organizations".

***B. Mainstream/scale up leasehold forestry into other modes of forestry, farming systems and community development***

B.1. Develop synergy with community forestry

- Delineate parts of community forestry for leasehold forestry, and allocate certain parts for short periods for value chain development. This would involve assessing the potential areas of forests, and the extent of degradation, for such allocation.
- Transfer the methods of participatory well-being ranking and identification of poor households to the CFUGs.
- Work with groups to develop land allocation and management strategies, and to choose appropriate products for value chain development.
- Develop mechanism for using part of the CFUG fund to manage the lease plot allocated.
- Develop a marketing/harvesting protocol and benefit-sharing mechanism between poor households and the CFUGs.
- Develop self-monitoring and evaluation system and update the contractual agreement between the poor households and the CFUGs on a regular basis.
- Pilot the intervention at selected district or CFUGs and document/disseminate the lessons learned.

**Integrating Community Forestry and Leasehold Forestry – Community-Based Forest Management**

Community and leasehold forestry are two modalities of what could be called Community-Based Forest Management (CBFM).

Integration would be possible if the government adopted a single CBFM policy that provided a level playing field for both approaches. This would require that amendments be made to the existing legislation and regulations, which tend to give priority to community forestry – notably, section 30 in the Forest Act and section 39 in the Forest Regulation would need to be amended so that they only refer to cases where leaseholds for purely commercial purposes are considered and not the CBFM leasehold approach.

While any part of the national forest can be handed over for community forestry, this does not apply to leasehold forestry for the poor, who are only entitled to lease degraded forests with a crown cover of 20 per cent or less. This is based on a regulation of the Ministry of Soil Conservation and is not based on legislation or regulations. This inequality would need to be reconsidered and the burden of forest recovery should not rest on the poor only.

Some Community Forestry User Groups have been entitled to use degraded parts of the community forest under an internal agreement between the group, which is reflected in the operational plan of the CFUG and endorsed by the CFUGs general assembly. [Editors note: This is guided by the Community Forestry guideline where inclusion of Leasehold Forestry concept in the Community Forestry was made possible with the name of "land allocation", by forming a sub group of poor families from within the Community Forestry members. This was piloted by the TA LFLP. A problem may arise, is that at renewal of the operational plan after every 5–10 years as stipulated, if the CFUG (possibly with new leadership) does not endorse it, then the sub-group has to forfeit the area. They may have invested much inputs to make it fertile and productive, however if they lose the benefit from it, it will be a hard blow to already poor community members. Thus tenural certainty is central.

A CFUG cannot issue leasehold contracts for long periods. To ensure a proper legal basis for such a long-term arrangement, the group would need to return their users' right of the degraded part of their community forest to the District Forestry Officer, who subsequently leases it out to the group of poor households. In this case, the group would surrender control over that part of its forest and the leasehold group would obtain users' rights for 40 years, which it may be reluctant to do (but on rare occasions has happened).

At present, leasehold forestry is administered in the Department by a unit, while community forestry is handled by a Division. A proposal was approved to create a single Division for Participatory Forest Management or CBFM, including two equal sections, one for Communautry Forestry and one for Leasehold Forestry. The proposal has been approved, and there is a Division looking after all of the CBFM. However, there are still separate units that have not been merged.

*Jaap Reijmerink (Consultant Economist, LFLP supervision missions)*

B.2. Develop and promote leasehold forestry in public land in Terai and Inner Terai

- Identify and delineate the public land, including roadside, canal banks and rivers, and abandoned flood plains in consultation with authorities, and develop models and benefit-sharing mechanisms for their productive use.
- Document and scale up best practices based on the existing management model of agroforestry or leasehold forestry established by BISEP-ST/MFSC and LFP/DFID.
- Provide material, and legal and technical support to establish public land forestry.
- Provide training to government line agencies on rehabilitating flood plains, stabilizing river banks, soil conservation and watershed management.

- Support the identification of beneficiary households and the development of a legal agreement with well-defined benefit-sharing mechanisms between the owners of the land and the lessees.

B.3. Promote leasehold forestry in protected area management systems

- Explore the concept of integrating leasehold forestry into biodiversity conservation and tourism development.
- Work with the government, especially the DoF, to ensure that there are mutually beneficial pro-poor strategies and incentive packages to scale up leasehold forestry in buffer zones and conservation areas.

B.4. Develop land use models and incentive packages for watershed conservation and carbon markets

- Given the significant contribution of leasehold forestry on watershed conservation, reducing carbon dioxide emissions and climate change mitigation, land use models that best capture these benefits will need to be developed based on two FAO studies that are currently underway with financial assistance by IFAD.

In all future endeavours, it will be important to develop orientation and capacity training packages for the staff of the different forestry institutions (e.g. protected areas authorities, soil conservation offices, forestry technicians), as well as forest user groups and their networks, and NGOs affiliated with the forestry sector. It will also be important to establish avenues or platforms for sharing learning and knowledge. These could include farmers' field schools, meetings at public places and schools, radio talk shows, etc.

**And let's not forget...**

The most significant change expressed by LFUG members was their ability to participate in making decisions about their future. This is particularly true among ethnic minority groups and *dalits*. Unfortunately, this aspect is often overlooked by those from higher castes, who so often dominate the social setting, as well as by community forestry committees, service providers and data collection/evaluation groups. This is why these groups should remain independent from community forestry committees. The original decision about who is poor enough to qualify for leasehold forestry can be made by the community, but it would be a huge mistake to make the leasehold forestry groups subordinate to the community forestry groups. Those who argue for a full merger either do not fully understand the social/power role implications that are playing out in the countryside or, if they do understand these, it is part of a political agenda (where the "party" takes care of "everything").

Since leasehold forestry groups are small and vulnerable, they can be easily intimidated and robbed of their assets. Therefore it is essential to have enough groups in a particular area and to link them to each other, to provide protection by numbers.

To increase the lease security of the leaseholders, it will be important to start paying a lease once the area is productive enough. In many places this could start after ten years. This would also strengthen the local fiscal situation (if the funds go to local government) and could strengthen local governance and decentralization.

*Frits Ohler (Former Chief Technical Advisor, FAO support to leasehold forestry)*

## Leasehold Forestry: lessons for institutionalization[28]

The goal of the Leasehold Forestry and Livestock Programme is to reduce poverty of 44,300 disadvantaged households by increasing production of the forest and livestock. Similarly, it aims to manage 31,000 hectares of forest lands, to conserve the environment, to develop the capacity of government institutions and to implement leasehold forestry as a poverty reduction programme in a gender sensitive way. To date, more than 20,450 ha of forest land has been leased to 38,436 households within 4,101 groups. The achievements are many, and some are described below.

**Environmental rejuvenation.** Biodiversity and vegetation cover increased from 32 per cent in new sites to 90 per cent after seven years. Fodder and forest products were harvested where little or none were produced before. The majority of leasehold forest plots are supplying forage and are fulfilling at least 25 per cent of annual fodder requirements of users. After five years, women are now saving 2.5 hours each day collecting fodder. About 69 per cent of forest land has been covered by various types of plants. Now, about 35 per cent of forage and fodder needs are supplied from the leasehold forests in LFUG member households.

**Poverty alleviation**. Before the programme, food security was very low, with a food deficit ranging between three and nine months. Now 11 per cent users have sufficient food for the whole year. Similarly, the average income has increased 70 per cent. Almost 90 per cent of households save money, at an average of NPRs 18 in monthly savings. In leasehold forestry households, goat herd size has increased (on average from two to five per household), with a shift to stall-feeding, while in non-leasehold forestry households herd size decreased. However, the services of Village Animal Health Workers have still been insufficient to the farmers.

---

[28] By Pashupati Nath Koirala, Forest Management Officer, Department of Forests, Nepal.

**Social empowerment.** The representation of women, *dalits*, *janajatis*, and the poorest in the key positions (Chairperson, Treasurer and Secretary) is: 36 per cent, 12 per cent, 32 per cent and 24 per cent, respectively – which is closely proportionate to the household composition of groups. The groups and members have been empowered to collect external resources to implement their Livelihood Improvement Plan.

## *The challenges ahead*

Undoubtedly, the Leasehold Forestry and Livestock Programme has been a success in rejuvenating degraded forest land and supporting poor rural people to improve their livelihood opportunities by empowering them socially and economically. However, several challenges remain.

### Priority of community over leasehold forestry

The Forest Act of 1993 has a weak arrangement of leasing land. There is no provision for "leasehold forest user group" in the Act or in the Forest Regulation of 1995. In community forestry, the "forest user group" has been recognized as an autonomous legal entity. Before handing over to the leasehold groups, the District Forest Office should provide a 35-day public notice with a provision to local people if they are willing to take a proposed leasehold forest as community forest. If the local people respond positively to the community forest, the forest is handed over as community forest rather than leasehold forest.

Similarly, the handover of leasehold forest to the poor household poses a serious question. In practice, the groups only get the "land", and for a specified period of time. The forest rule stipulates that the lessees have to keep the trees until the government utilizes them. It has created extra responsibilities, and costs, for poor households to protect the trees of the government without any extra benefits for doing so.

## Inheritance right to membership

The right of the lessee to transfer his/her lease to someone else in the event that the lessee dies or migrates from the area is one of the policy issues. The regulation has a provision for a lessee to transfer or sell the lease to someone else under two conditions: if the lease period has passed at least one-third with a "satisfactory" implementation of the operational plan and if the Ministry grants prior approval. In the case of community-based leasehold forestry, it is a complicated task because the lease is granted to the "group" and not to an individual. The likelihood that the entire group would sell or transfer its lease to another group is almost nil. A legal provision allowing an individual member of a group to sell the lease to another poor household or transfer to another member of the same household would mitigate the issue. Prior approval of the Ministry is also not practical.

## "Project" status of leasehold forestry

The provision envisioned in the Leasehold Forest Policy of 2002 to be developed in a programme approach has not yet come into effect. Although, the LFLP is said to be a programme approach, transformation of the policy has not been institutionalized through including it in the Forest Act and Regulation. As a result, it is still in tentative status. It is only possible to execute the programme with developing a programme for poor people. At the same time, other forestry projects are incorporating the concepts into their programme, such as the Multi Stakeholder Forestry Programme and Community Forestry Development Programme, but their modalities have not followed the core full components for convergence towards poverty alleviation of the poorest community as in Livelihood and Forestry Programme.

## Policy recommendations

The government has accepted a valid programme for disadvantaged people, particularly in the forestry sector, to address poverty-related problems in rural areas. To run this programme regularly in all districts and through government agencies, a number of rules and policies of the Forest Act will need to be amended. For example:

- The Forest Act of 1993 should give equal priority to community and leasehold forestry. After disadvantaged people in rural areas have gained an equal footing, they will have access to usable forest land, which will enable them to increase productivity and improve their livelihoods.
- At least 50 per cent of the benefits should be allocated to leasehold forest users for recorded standing trees of the forest. Lessees are reluctant to conserve the large trees on their leased territory. As a result, a sense of ownership has not been sufficiently internalized to conserve the existing tree species. Instead, users focus their efforts on collecting and cultivating the grass species.
- Inheritance of leasehold members should be clearly defined in the Act. The policy should spell out the position of the leasehold forest after 40 years of leasing. This will promote continuity of forest management and conservation. At the same time, it will reduce conflict among potential users of the leasehold forests.
- The government's own approach should be not to have differences among different types of leasehold forestry. If it is established as a primary programme in the forest policy or Act, then District Forest Offices and forestry professionals will implement this kind of forestry regime as community forestry and the LFUGs will obtain a legal and institutional status that is equal to other forestry regimes. Thus, it is recommended that this programme be a regular component of forestry, as a continued programme approach, in order to have an impact on the country as a whole.

It has been proposed by the Department of Forests and the Ministry of Forest and Soil Conservation that all kinds of forestry programmes be put on an equal footing. This Bill was already in the Ministry of Law and Justice for review, but it bounced back to the Ministry of Forests, i.e. it was rejected. Only when there are elections and a constitution drafted will it be possible to try to have the Bill passed.

**Keys to success**

Capacity is the key. If the farmers can farm very well, they will begin to imagine other scenarios and expand their livelihoods. If their desires can correspond to their capacities, great progress can be made. But if, for example, they are not interested in having goat or cattle and we provide it, they will do nothing. Their desire, their ambition and their interest must be there.

Inter-agency coordination is also a key. Leasehold forestry needs to be a leading initiative in all contexts. Donors, media and government officers need to be aware and involved, and convinced that leasehold forestry enhances the lives of people who essentially have nothing.

And then there is transparency, without which the initiative will not work. If we are open and frank and don't think in hierarchical terms, things can work. It is so much a function of transparency and simplicity. These are the main tools of cooperation. If those in charge are transparent and simple, coordination will take care of itself.

*Dr Karki Ram Kumar, Senior Veterinary Officer, District Livestock Service Office, Chitwan*

**Capacity all around**

There are many organizations involved that are important in making the programme a success. However, so many people involved in the programme need capacity development. The District Education

Offices have several programmes for formal and informal education. Education is the only thing that helps. Once people are capacitated, they can continue with income-generating activities.

Many people are illiterate and have little education. They need more capacity development activities and *regular* support – for example, in holding monthly meetings, keeping their records, completing forms, and so on. There are thousands of groups, and each group has about ten members. Add it up and it is a lot of people. We have about one social mobilizer for every 170 people. Moreover, many forestry and livestock staff have little knowledge of rural finance or finance management. It's an enormous task and a big challenge for sustainability.

*Mr. Biswa Nath Oli, Director-General of the Department of Forests under the Ministry of Forests*

**Yes we can**

I am fully aware of the termination of the project. A lot of planning needs to be done. Many more cooperatives should be established. All of the staff are visionary, they are dedicated, they are present. The hub office is always there, watching, observing, hoping. The government has to be the vehicle for promoting leasehold forestry. I am sure it can be done. And it must be done. I am very hopeful.

*Mr Jhapendra Bahadur GC, former Chairperson of the District Development Committee, Palpa*

## Integrating Leasehold and Community Forestry[29]

Leasehold forestry and community forestry are complementary programmes that can function together in a mutually beneficial

[29] By Kaushal Shrestha, environment specialist IFAD country office.

system. A number of case studies can illustrate the success of integration.

- ***Allocating/converting degraded land within community forests to leasehold forest***

Cases exist where community forest users themselves have taken the initiative to allocate barren land within their community forests to poor members of the community. Therefore, while not officially in the system of leasehold forestry, the integrated mechanism of leasehold and community forestry co-existing within the same system is already functioning in various project areas. Cases also exist (in Lamjung) where, during renewal of the lease, degraded/barren areas within community forests were officially converted into leasehold forests, and then subsequently provided to poor members of the community on a 40-year lease. This was achieved through cooperation between all concerned parties.

Examples of pro-poor provisions in CFUG constitutions and operational plans

- Subsidies in prices of forest products.
- Reservation of spots for the poor, women Dalits (untouchables) on community forest committees and decision-making bodies.
- Special provisions for the distribution of forest products to vulnerable groups (for example, charcoal to blacksmiths, products freely distributed to victims of natural disasters, single women, or conflict victims).
- Allocation of CFUG funds and low-interest loans for income-generating activities.
- Forest resource management with allocation of Community Forestry Programme land to poor users.
- Scholarships to children from poor families.

- ***Leasehold forests as a buffer to community forests***

A case in Shaktikhor (Chitwan) exists where the leasehold forest surrounds the community forest, thereby acting as a buffer zone. The surrounding leasehold forest protects the ecosystem within the community forest from any human disturbance, and also acts as a link between other patches of community forests for animals to migrate, thereby promoting biodiversity. Such a structure can be replicated during the allocation of forests.

- ***Leasehold forestry as Phase 1 towards (reformed) community forestry***

Following rehabilitation of forestland and poverty alleviation of users, inner areas of leasehold forests can be converted to community forests. Various leasehold forests have been rehabilitated in the past few decades, the users of which have also lifted themselves up from poverty through support from leasehold forestry projects and schemes. A good example has taken place in Rayale (Kavre) and Kurintaar (Chitwan). Thus, having achieved its goal, there remains little reason to continue the leasehold forestry programme for the remainder of the 40-year lease. Subsequent to forest rehabilitation and poverty alleviation (Phase 1), inner regions can be converted to community forests, applying (revised) regulations to promote core density and biodiversity as well as to increase revenue from forest products for both user groups and the government. In this regard, regulations must be revised accordingly in order to provide supportive environments for integration. In addition, schemes to encourage LFUG members to support conversion will be an important factor in the successful implementation of this idea.

**Approach towards integration**

**Local intervention (bottom-up)** – The fact that community forest users and other local decision makers themselves took the initiative to convert community into leasehold forestland shows that an environment already exists at the ground level where leasehold and community forestry can be integrated without creating conflicting situations between the two user groups.

**Policy intervention (top-down)** – The case of Lamjung implies that there exists a policy within the community forestry framework where land allocation can be altered, and suitable areas within community forests can be converted into leasehold forestland.

**Policy implementation** – The cases in Chitwan and Lamjung suggest there is a grassroots need for a strong policy framework to support integration of leasehold and community forestry. Therefore, policies must be created, and in case they already exist, the Ministry of Forestry must implement them. Additional support and resources, human and financial, must be allocated to strengthen and accelerate this process.

## Kathmandu Declaration 2014 on Pro-poor Leasehold Forestry
## LFLP closing workshop, 11–13 June 2014[30]

*Respecting the active participation of national and international participants associated with this programme, leasehold forestry users, well-wishers, intellectuals, governmental and nongovernmental staffs, social mobilisers, experts and donor agencies, the regional workshop on pro-poor Leasehold Forestry held in Kathmandu from 11 to 13 June, 2014 proclaims this Kathmandu Declaration – [editor's summary:]*

*1. Follow through on the relevant amendment on Forest Act-2049 which has been proposed, enforcing similar legal priority to all forms of the community-based forest management systems.*

[30] "Leasehold Forestry: A New Dimension on Livelihoods" – Regional Workshop on Pro-poor Leasehold Forestry, 11–13 June 2014. Workshop Proceedings. Department of Forests, Ministry of Forests and Soil Conservation, Nepal, and FAO Nepal.

*2. Revisit the provision that restricts the leasing forests with only up to 20% of the crown cover.*

*3. Ensure equitable benefit-sharing mechanism on old trees present at the time of the leasehold forest hand over.*

*4. Appropriate legal provision to protect the rights and interests of the leasee is demanded as the concept of leasehold forestry is extended to community forest and public land management.*

*5. Provide strong commitment from all development partners to take Leasehold Forest User Groups as the entry points for all activities aimed at poverty reduction at rural set-ups, including implementing livelihood improvements.*

*6. The Silvo-Pastoral agro-forestry model has been shown to rehabilitate forests and control land degradation – its contribution to enhance climate change risk minimization programmes/activities should be increased.*

*7. The integrated forestry and livestock development extension approach should also be continued in other districts in order to provide additional contribution to overall poverty alleviation.*

*8. The development of specialised professional resource persons at the local level is vital for the mobilization of the leasehold forest user groups.*

*9. Future extension of the programme should consider the high potential of leasehold forestry at the landscape level.*

*10. Publicly promote the positive contributions the Leasehold Forestry programme.*

*11. To organize another leasehold forestry workshop within five years.*

**13 June 2014 (30 Jestha, 2071 B.S.), Kathmandu, Nepal**
(Unofficial Translation)

**Kathmandu Declaration signed by stakeholders on 13 June 2014**

# *Chapter 7*

# Conclusions

These pages have described more than 20 years of history of leasehold forestry: how it evolved from an initial government recognition of the interrelated problem of poverty among the poorest forest users and severely degraded forest lands into a programme that expanded from few communities to a national scale, learning many lessons on the way, to establish new integrated modes of tackling deep-rooted issues. During this journey, it has developed important assets for poor rural people and strengthened land tenure and security. It has also fostered community action and inspired similar initiatives in related forest environments. Some of the central tenets of the programme have valuable lessons for other forestry and livelihood programmes in Nepal and beyond.

To achieve these landmarks, the leasehold forest approach has taken up a forest management strategy that is radically different from traditional ones. In particular, management and benefits are based on non-timber products, and a mix of plantations with grass/forage and trees, which address the broader requirements of the marginal users. The forage/grasses provide the life-line of the users, who in turn conserve tree species with the hope that they will provide them with their livelihood-sustaining fodder, fuelwood, bark, leaves and litter. The timber is only a secondary benefit. Finally stall-fed livestock and goats come as an additional income to the family. Such an approach contributes to a marriage between livestock and forests and the often otherwise antagonistic goals of poverty alleviation and ecosystem improvement.

The projects have also demonstrated that poor users of leasehold forests are not the culprits of forest degradation, as the development discourse so often espouses. In fact, with an introduction of compatible management models that incorporates their livelihood strategies, they are also friends of nature. And they are dependable and bankable.

The leasehold approach of Nepal has essentially been one in which the government placed some of the "cost of turning the degraded forests into greenery" on to the poor, but at the same time compensated a job well done with incentives – in the form of dependable tenureship of land, training and inputs. Early on, it was recognized that farmers starting with a very low asset base needed their surplus labour to accumulate into some greater tangible short-term benefit – in this case livestock (normally goats), which can be exchanged for money whenever necessary. Thus the strategy has almost counter-intuitively added a livestock development approach, deeply linked to the traditional forest user system – and often traditionally the causes of forest degradation – as a boost to farmers' livelihoods and as an incentive toward better forest management.

The limitations of the model must also be taken into account. When the model works well, it will yield a sum equivalent to NPRs 300,000 (about USD 3,000) per year per household for almost 20 years. This is significant additional income for typical marginal Nepal hill households, but it cannot be expected to yield drastically higher returns. Thus, this form of management is feasible only in areas where the level of income is thought to be substantive and forms a part of the livelihood strategy. If less land is available per household, or livelihood models with fewer returns are used, it could exacerbate poverty rather than alleviate it. Essentially there are two prerequisites: surplus labour and surplus degraded forest land around the community dwellings. Therefore, the approach could be replicable in some parts of South Asia, East Asia, Africa and South America, but should not be thought of as a panacea for all conditions of entrenched poverty. The leasehold forestry story narrated in these pages also speaks of some of the considerable challenges in putting in place and scaling up this model: challenging stereotypes, building awareness, working out appropriate procedures, building capacity on a large scale, and pushing for policy shifts. It is not an easy path, and those wanting to go down it must be prepared to invest much time and effort and commitment.

Leasehold forestry and the intense engagement by communities, the forestry department and officers, agencies, policy makers, and technical agencies and donors, has brought with it some important conceptual and practical lessons – and probably many more are to come. As part of an ongoing wider dialogue we offer a sample:

**Addressing forest livelihoods of the poor with a multifaceted approach and the communities' own institutions.** Poverty alleviation needs to be thought of as fundamentally having to address the multifaceted management of household livelihoods. Poor rural people have their own microcosm and, accordingly, poverty has to be tackled at a micro level, not a macro level. Every household has its own livelihood strategy, recognizing and expanding its sphere of action through tools like individual and group livelihood improvement plans. Poverty is a multi-dimensional and multi-level phenomenon, and it is always difficult to disentangle the causes and results. The mixed results of conventional poverty alleviation strategies point out that the strategies employed to combat poverty have failed to take into account the poverty process, not uncovering the multi-dimensional causes and factors of poverty. Previously, accelerating economic growth by investing in projects with high returns was thought to be the cornerstone of poverty alleviation strategies, and it was believed that the inevitable result of economic growth is poverty alleviation. Projects were conceived to transfer technology, provide subsidies, and create infrastructure, and it was hoped that poverty reduction would follow. It is clear in Nepali and wider experiences, with pockets of poverty deeply rooted in settings of fast rising GDP, that economic growth alone cannot reduce poverty without dedicated strategies to the most vulnerable people.

Thus, leasehold forestry provides an alternative vision of development that centres on enlarging people's choices and capabilities, and provides for their participation in decisions affecting their lives. By putting users and their groups at the centre of the process, leasehold forestry has developed a system of governance that promotes and supports users to articulate their interests, exercise their rights, meet their obligations, and mediate their differences. This is a true contribution to a governance that promotes participatory decision-making and transparency of action in all spheres of life. Local-level governance through local institutions enables and empowers people to participate more directly in decision-making, is in a position to produce quick responses to people's needs and priorities, and is one of the critical ingredients for promoting ownership by the people. What has been learned is that the poor must be encouraged to build their own institutions, where they can mediate differences, discuss their livelihoods, explore in their own way their way of life, and make decisions affecting their lives – the best antidote to powerlessness. This is exactly what the local leasehold user groups have strived towards. The small (5–15 households) and homogenous (all poor) group composition, their like-mindedness and their sharing of similar difficulties have pushed them to collectively achieve what the larger populace seldom is able to.

**Addressing social exclusion and empowerment.** Effective targeting and real engagement of the traditionally marginalized is the next step. Instead of assuming the poor to be passive beneficiaries, it is necessary to reorient thinking to consider them to be active partners, focusing resources on the poor directly – including helping them build their own institutions, and with appropriate institutional incentives, and giving them a say in how resources are allocated and what types of service delivery mechanisms are put in place. This does imply further requirements.

As we have noted before, the understanding of a comprehensive approach, i.e. livelihood security that goes beyond employment generation or pure forest management, is a must. This is particularly important in order to be inclusive, and for users to respond and benefit. Given the right opportunities, the rural poor have considerable potential and capacity to contribute to the national economy, as well as improve their own standards of living. They can manage forest rehabilitation and infrastructure development, they can mobilize savings. They are “bankable” and “trainable” and ready to engage, grow and take over. There are returns on investing in them.

Going beyond simple income deprivation, social exclusion is a state of poverty in which individuals cannot access the living conditions that would enable them to both satisfy their essential needs and participate in the development of the society to which they belong. Therefore, when people cannot achieve their potential by upgrading their capabilities because of deliberate and structural constraints (such as caste, ethnicity, religious orientation, gender or other social barriers), the exclusionary processes that remain are the major obstacle to poverty alleviation. Such exclusion has been pervasive in Nepal and South Asia, often typical of more isolated natural resource-dependent communities. Participatory development, the empowerment of local communities, and the devolution of authority become nullified if social exclusion holds sway and social inclusionary principles are not deliberately fostered. The leasehold forestry experiences have shown that inclusive processes and institutions are possible in a wider technical development framework, and can be further fostered. There should be wider recognition of a comprehensive developmental framework that affirms institutions and human and social capital as no less critical than physical and financial capital in the process of change from despair and deprivation to development and human well-being.

**Further building core capacities.** Sustaining and scaling up such locally focused and empowering programmes always is a challenge. There needs to be continuous monitoring and preparedness when it comes to the transfer of technical know-how to user communities. It cannot be done by government departments or NGOs alone. Under leasehold forestry it has been amply shown that social mobilization activities are a driving force for success. Local community members and users, those knowledgeable about their own systems and needs, and the development of Local Resource Persons in both technical support and monitoring are the true eyes and ears of the process. There is more to do here, and to internalize this on a long-term basis in future leasehold forestry initiatives.

Capacity of communities must include critical additional development mechanisms. The poor are vulnerable to many kinds and magnitudes of external shocks. Vulnerability and coping mechanisms are therefore crucial for long-term sustainability. They can be introduced in the form of Rural Finance. The leasehold experience has shown that member-based and -owned Rural Finance institutions are critical for leveraging further opportunities and resources. As we have seen, this is still an unfinished chapter of the leasehold story, and will need sustained links to other dedicated rural finance programmes, a challenging topic in any context.

Leasehold forestry has shown how important it is to build on the communities' own strengths and existing livelihoods systems. In so many rural settings, introducing new things may not only create doubts but also create scepticism towards the change. Change should be carefully brought in, evolving on existing experiences and traditional specializations, and fitting local economic potential. Goat raising has a long tradition in Nepal, and proved to be an ideal "entry point" for the vehicle of development, with other activities having revolved around it. The next step to develop core economic activities, such as goat production, at more sophisticated marketing and technical levels, is ready for the taking as demonstrated in the oldest leasehold groups (Kavre district, for example, where many LFUGs turned into cooperatives: dairy production, fish, vegetables, etc.).

With almost 40 per cent of the land under forest, the forestry sector has a great responsibility towards the poverty reduction agenda. It can neither allow the forest area to be reduced nor shirk from the main responsibility of poverty alleviation. This is where the leasehold forestry concept comes into play. Its potential is great in terms of getting the community forest management process right, and providing additional resources to the people who need it most and who are highly dependent on forests. And, perhaps most importantly, it has the capacity to both enhance the environment and reduce poverty together, without jeopardizing one or the other.

As this book on LFLP and its 20 years of experience in poverty alleviation and forestry development comes to an end, we must remember that the many lives discussed in these pages will continue long after the book has been put on the shelf, as will the generations that follow. In this year 2015, which promises to bring forth a number of worldwide initiatives on climate change and environmental protection, we hope that this story from the remote Himalayan foothills and people of Nepal will bring some hope – and some solutions – for the future health of the planet.

# About the authors

**Brett Shapiro** (brettjshapiro@gmail.com) is a writer and journalist. For the past 20 years, he has written and edited technical, policy, strategy, advocacy and general audience publications for more than a dozen United Nations agencies. His United Nations work also includes two full-length books: *Nourishing the Land, Nourishing the People: The Upper Mandrare Basin Development Project in Madagascar* (published by L'Harmattan in 2008 and CABI in 2010) and *PARECAM, From Food Crisis to Food Security: One Programme in Madagascar Makes a Difference* (under publication). He also provides specialized training in writing to UN staff at the United Nations System Staff College in Turin, in headquarters offices and in the field, as well as to diplomats at the Italian Ministry of Foreign Affairs. In addition, he is a visiting professor at the Università di Siena.

Outside of his UN and development work, Mr. Shapiro is the author of the best-selling *L'Intruso*, a memoir published in Italy (Feltrinelli) and subsequently a film and theatrical production. He collaborated with the conductor Robert Craft on a biography of Igor Stravinsky entitled *Dearest Bubushkin*, published in the United Kingdom (Thames and Hudson). He has also written two children's books, *The Adventures of Bailey the Bag Dog* (1998), winner of Austria's National Book Award and published in France, Germany, Austria and South Korea, and *Bailey Strikes It Rich*. His essays and articles have appeared in magazines and newspapers in Italy and the United States, and he has made many guest appearances on Italian television.

**Govinda Prasad Kafley** (gpkafley@gmail.com), a Nepali national, is a freelancer forest professional. He obtained his Masters in Forest Resource Management from the Graduate School, University of the Philippines at Los Banos in 1992. After completing an Associate of Indian Forest College (AIFC) at Dehra Dun, India, he joined the forest service of the Government of Nepal in 1981 and worked in different professional capacities at the Ministry of Forests and Soil Conservation (MFSC) of Nepal until 2009. Before leaving MFSC, he worked as Division chief, Environment Division MFSC, Regional Director of Forests and Far Western Development Region, Nepal. During 2003–2008, he was a Programme Coordinator of the Leasehold Forestry and Livestock Programme (LFLP), an IFAD-funded Government of Nepal programme, which was exclusively meant for poverty alleviation through Natural Resource Management. During 2004–2006, he also worked part time as a Liaison Officer of the Indonesia-based Center for Forestry Research (CIFOR). After leaving the government service, he joined the FAO to lead the Technical Assistance Project for Leasehold Forestry and Livestock Programme (1 December 2009–16 July 2014) as a team leader. Now he is active in forest professional activities and is involved in master planning process of Chure region of Nepal as a natural resource economist.

**Jim Hancock** (jim.hancock@fao.org) is a natural resources management and livelihoods officer with the FAO Investment Centre in Rome. Jim has a BSc in Environmental Biology and MSc in Aquatic Resources Management (University of London, UK). Jim has over 20 years of experience in natural resources management, community-driven development, rural institutions, and livelihoods of marginalized fishers and farmers, and landless agricultural households. Before joining the staff of FAO in 2008, Jim worked in fisheries research in Chile, 7 years in coastal management and community development in the Philippines (AsDB and EU projects), as senior scientist at the Natural Resources Institute, UK (DFID projects), and as a consultant with FAO and IFAD on programme design in Asia, emergency recovery (including avian influenza and tsunami), and Word Bank strategies on scaling up. Since 2008 with the FAO Investment Centre he has led and contributed to a number of projects design, supervision, and evaluation missions, largely focusing on natural resources and livelihoods. Much of the work has been in India, but also Sri Lanka, Bangladesh, Nepal, Mongolia, China, Vietnam, Pakistan, Indonesia, and the Pacific Islands. The programmes and projects cover women's and village livelihoods development, smallholder land–water management, coastal and small-scale fisheries management, forestry and pastoralism. He led the team for programming FAO support after Typhoon Haiyan in the Philippines. He has a particular interest in participation by communities in decision-making and the development process. He is working with the Governments of India and Nepal and IFAD on assessing project experiences for scaling up programme lessons and contributed to various guidance on monitoring and evaluation.

**Bala Ram Adhikari** (adhikari.balaram@yahoo.com) is a professional forester in Nepal. He received his forestry degree from the Indian Forest College Dehradun and did his MSc in Forestry Extension at the University of Reading, UK. He has 30 years of experience in the field of forestry. During his public service period, he was heavily involved in supporting policy development and implementation of the forestry sector and governance build up. He worked as a Programme Coordinator for the Leasehold Forestry and Livestock Programme for about six years (November 2008–August 2014). He is currently working as a Regional Director for the Far Western Regional Forest Directorate, Dhangadhi, Nepal.

**In memoriam: Jacob (Jaap) Reijmerink (1943–2014)** was an economist and team leader with extensive agriculture and rural development project experience in a wide range of agencies and countries. Jaap did a Master's degree at the Free University, Amsterdam (1971) and a course in Agricultural economics (1981). From 1973 to 1981 he was advising, planning and evaluating projects on employment and labour intensive public works programmes with ILO and UNDP, much of it in West Africa. From 1982 to 1987 Jaap was Project Controller in the Africa Division of the International Fund for Agricultural Development (IFAD), Rome. In this role he initiated agricultural development projects for smallholders with a poverty alleviation focus, and supervised and monitored implementation in four African countries: Senegal, Mali, Niger and Ethiopia. From 1987 onwards he worked as an independent consultant in agricultural economics, agricultural credit and rural development. He did so for the World Bank, the International Service for National Agricultural Research in The Hague, the Dutch Ministry of Development Cooperation and, most often, for IFAD in Rome. He led many missions for different stages of project design and implementation, including strategy formulation, inception, formulation, appraisal, evaluation, supervision and completion analysis. Missions were fielded in 10 countries in Asia, 18 countries in Africa, as well as North Africa. His expertise covered a broad range of topics, including: macro-economic and socio-economic analysis, target group definition and participatory methods of project management, credit, micro-finance and village banks, biological agriculture, community development, institutional analysis, management and organization, programming of implementation as well as overall project coherence, gender and poverty alleviation orientation. In Nepal he led most of the Leasehold Forestry and Livestock Programme supervision missions, and was very fond of Nepal and its people. He will be missed by friends and colleagues in Nepal.

**Pashupati Nath Koirala** (koiralapn@gmail.com) is a professional forester in Nepal. He has 28 years of experience in forest and natural resources management. He is heavily involved in supporting policy development and implementation in the forestry sectors and governance build up. He has written several research papers for international and national conferences. His interests and writing focus on livelihoods support of the poorest communities in Nepal. His major contributions in livelihoods include 'Community movement to save land for their livelihoods support (a story case of a local community to maintain land degradation on the stream banks of Nepal)', a poster presented at the International Disaster and Risk Conference (IDRC), Davos, Switzerland in 2014; 'Environmental entitlement and benefit sharing in community forests in Nepal' and 'Degraded forest management through leasehold forestry for poverty alleviation in Nepal', a paper and a poster presented at the UNCCD 2nd Scientific Conference at Bonn, Germany in 2013; and 'Livelihood improvement through degraded forest management in Nepal', a paper presented at the International Disaster and Risk Conference in 2012.

His contribution in Department of Forests as a Forest Management Officer and Planning Officer was used in Leasehold

Forestry and Livestock Programme. He received an award from the Joint Japan World Bank Graduate Scholarship Programme to study for a two-year Master's degree at Wageningen University in the Netherlands.

**Kenichi Shono** (Kenichi.Shono@fao.org) is currently working as Forest Resources Officer at the FAO Regional Office for Asia and the Pacific. He has broad experience in forest management and conservation, including project coordination, research and policy analysis, with extensive field experience in Southeast Asia, Central America and the USA. He first gained experience as a field researcher conducting an ecological study on big-leaf mahogany in Central America for the Center for International Forestry Research (CIFOR). Subsequently, he coordinated a research project on forest restoration in Singapore for the Smithsonian Tropical Research Institute. From 2005 to 2008, he worked as Associate Professional Officer at FAO where he was involved in various programmes and projects related to promoting sustainable forest management in the Asia-Pacific. From 2008 to 2013, He worked for PT Hatfield Indonesia as Senior Forestry Specialist and Technical & Operations Director. In addition to fulfilling his corporate management role, he provided technical inputs to a range of forestry-related projects, including: forest management certification; biodiversity assessment and management; high conservation value assessments; forest restoration; carbon stock assessments; REDD+ project feasibility studies; and land use carbon project certification. In 2013, he re-joined FAO in his current post.

**Nav Raj Baral** (baralnavraj1@gmail.com) is a forestry and environmental management specialist in Nepal. He graduated in Forestry from Indian Forest College, Dehradun (1982) and in Environmental Management and Development from the Australian National University, Canberra, Australia (1995). He worked for the government of Nepal for about 23 years (1982–2004) and has over 20 years of experience in policy formulation, strategic programme development, planning and management in the field of community natural resources (land, forests, pasture, and water). As a forestry expert he worked for FAO, Nepal in developing the Agriculture Development Strategy (ADS) 2013 and as a supporting team member in developing Forestry Sector Strategy (FSS) 2014. He has over 15 years of involvement and demonstrated expertise in programme and project reviews, assessments and evaluation as well as human resources development and capacity strengthening of community-based institutions/organizations at various levels. He has written two books (*Regeneration Management in Nepal* (1990) and *Forest Fire Management in Nepal* (2000)) and dozens of articles on forests, resource tenure, biodiversity conservation and community development. He has extensive work experience with international experts and multidisciplinary teams, with a wide spectrum of stakeholders including governments, NGOs and multi and bi-lateral development agencies including FAO, UNDP, IFAD, AusAID, USAID, and SDC.

**Benoît Thierry** (b.thierry@ifad.org) is an agro-economist based in the IFAD South East Asia Hub. He graduated as Engineer in Tropical Agricultural Economy from ISTOM, France (1987) and completed a Doctoral Programme in Human Geography at the Sorbonne University, France (1988). In the late 1980s, he was Project Manager in Bolivia and in Western Mali. In the 1990s, Benoît was Representative for West Africa of GRDR, an NGO specialised in remittances. From 1995, as Project Manager in Cambodia with GRET he managed a rural water and sanitation programme. Then for UNDP Cambodia he monitored the CARERE reconstruction programme from 1996 to 1999. From 2000, he was Portfolio Manager with UNOPS in Kenya. Since Benoît joined IFAD headquarters in 2004, he has been the country programme manager successively in charge of Rwanda, Zimbabwe, Comoros and Madagascar. In October 2011, he moved to the Asia division where he was in charge of Bhutan, Nepal and Thailand, and supported the Farmers Organisations network in the Asia-Pacific. Since September 2014, he has been based in Hanoi to open the South East Asia Hub of IFAD. Throughout his various assignments, and convinced of the needs for better information about rural development, Benoît created knowledge management products, from the GRDR videos and exhibits to the CARERE website, and initiated an coordinated the teams on the Madagascar book *Nourishing the Land, Nourishing the People* (L'harmattan 2008, CABI 2010) and the present volume *Regenerating Forests and Livelihoods in Nepal: A new lease on life*.

## List of Leasehold Forestry and Livestock Programme Technical Assistance project publications

| *Title* | *Year* | *Remarks* |
|---|---|---|
| Potential Agro-forestry Practices in Leasehold Forests and Shifting Cultivation Areas for Land Rehabilitation and Socioeconomic Upliftment of Local People in Tanahun District | 2011 | Outsourced |
| Study on Optimal Size of Goat Herd for Leasehold Family | 2012 | Outsourced |
| Review Assessment of Leasehold Cooperatives | 2012 | Outsourced |
| Kipat Land Tenure System in Eastern Hills of Nepal. | 2013 | Outsourced |
| Assessment of Carbon Sequestration Potential of Leasehold Forests: A Study of Katakuti village development committee, Dolakha District | 2012 | Master's Thesis |
| Assessing the Impact of Leasehold Forestry on Income Poverty in Nepal | 2011 | Master's Thesis |
| Assessment of Carbon Stock in Leasehold Forests of Two Sub-watersheds in Lamjung and Tanahun Districts | 2013 | Collaboration with ICIMOD |
| A Review of Marketable Non Timber Forest Products | 2012 | Project internal study report |
| Stratification of LFUGs: An Analysis of Group Performance | 2011/12 | Project internal study report |
| An Assessment of Outcome of LFLP | 2011/12 | Project internal study report |
| Results and Impact Management System (RIMS) Survey | 2012 | Project internal study report |
| Review Report of Goat Cooperatives | 2011 | Project internal study report |
| Gender Assessment and Action Plan (GAP) Preparation | 2011 | Project internal study report |

| | | |
|---|---|---|
| Baseline Study: Cluster Level for LFLP Districts | 2011 | Project internal study report |
| Draft Provisions to be Included in Buffer Zone Regulation (submitted to Department of National Parks and Wildlife Conservation) | 2011 | Project internal study report |
| Forage Development Package: A Hand Book to Farmers | 2012 | Project internal study report |
| Livelihood Improvement Plan (LIP) Implementation Guidelines | 2013 | Project internal study report |
| Conducting Participatory Monitoring and Evaluation at Group Level: A Hand Book for Field Staff | 2011 | Project internal study report |
| Conducting Joint Monitoring at District Level: A Hand Book for District Line Agencies | 2011 | Project internal study report |
| Implementation Guideline for District Level Coordination Workshop | 2012 | Project internal study report |
| Implementation Guideline for Field Level Interaction Workshop | 2012 | Project internal study report |
| Report on "Piloting the Demonstration of Portable Goat Dipping Tank" | 2013 | Project internal study report |
| Proceedings of Regional Workshop of Leasehold Forestry (LF) | 2014 | Project internal report |

# References

1. Albano, A., Kumar, C. and Regmi, B. (undated). Improving the livelihoods of the poor and marginalized in Nepal through Leasehold Forestry: A review of institutional constraints and opportunities. Final technical report. Local Initiatives for Biodiversity, Research and Development (LI-BIRD)

2. Baral, Jagadish C. (PhD), May 2004. Miscellaneous Paper Series, Western Regional Forestry Directorate, Pokhara

3. Baral, Nav Raj, April 2013. Review of Literature on Impacts of Leasehold Forestry (UTF/NEP/072/NEP) (draft)

3a. Bhatta, L.D., Gyawali, N., Lal Karna, A. and Shrestha, G.B., April 2012. Decentralised Forest Sector Governance. Prospects and Learning from Central Terai Region, Nepal. Netherlands Development Organisation (SNV), Nepal

3b. Bockel, L. and Grewer, U. Lessons learned from appraising the impact of IFAD projects in Nepal on the GHG balance and stocks of natural resources: an analysis using the Ex-Ante Carbon balance Tool. FAO, Rome

4. Department of Forests, report, 16 July 2014

4a. DFID, Nepal, December 2008. Seven years of the Livelihoods and Forestry Programme: Enhancing rural livelihoods through forestry in Nepal. Contributions and achievements

5. Department of Forests, November 2012. Leasehold Forestry and Livestock Programme. Annual Report 2011/12

6. Duquesne, Louise. ISTOM. 17 May 2013. LFLP Success Stories in Kavre District

7. FAO, 2012. Key Progress: Inputs and Achievement 2010–12. FAO Technical Assistance for Leasehold Forestry and Livestock Programme

8. FAO, November, 2012. Stratification of Leasehold Forest User Groups: An assessment of the group performance (TA for LFLP, GCP/NEP/062/FIN)

9. FAO, November 2012. Stratification of Leasehold Forest User Groups: An analysis of group performance (TA for LFLP, GCP/NEP/062/FIN)

10. FAO/IFAD, December 2012. Technical Assistance for Leasehold Forestry and Livestock Programme. Key progress 2010–2012 (GCP/NEP/062/FIN)

11. Government of Nepal, 2012. Multi Stakeholder Forestry Programme (brochure). See MSFP website: http://www.msfp.org.np/

12. IFAD, 6 December 2001. Report and Recommendation of the President to the Executive Board on a Proposed Financial Assistance to the Kingdom of Nepal for the Western Uplands Poverty Alleviation Project (EB 2001/74/R.19/Rev.1)

13. IFAD, December 2003. Kingdom of Nepal, Hills Leasehold Forestry and Forage Development Project (Interim Evaluation), Report N° 1431-NP

14. IFAD, December 2004. Report and Recommendation of the President of IFAD to the Executive Board on a Proposed Financial Assistance to the Kingdom of Nepal for the Leasehold Forestry and Livestock Programme

15. IFAD, December 2007. Making a Difference in Asia and the Pacific, Issue 18.

16. IFAD/Federal Democratic Republic of Nepal, March–April 2012. Leasehold and Community forestry, two approaches of forestry conservation in Nepal (Case study)

17. IFAD, May 2012. Leasehold Forestry and Livestock Programme, Joint Review Mission Report (No. 2668-NP)

18. IFAD/Federal Democratic Republic of Nepal, June 2012. Leasehold and Community forestry, two approaches of forestry conservation in Nepal (Press article)

19. IFAD, January 2013. Enabling poor rural people to overcome poverty in Nepal (booklet)

20. IFAD, April 2013. 35 Years In Nepal (booklet, ISBN 978-92-9072-4575)

21. IFAD, May 2013. Leasehold for Poverty Reduction in Nepal

21a. Ministry of Forests and Soil Conservation, 2005. BISEP-ST Annual Progress Report

22. IFAD website; last checked 25 June 2013

23. Ministry of Forests and Soil Conservation. Asia-Pacific Forestry Sector Outlook Study II. Working Paper Series. Working Paper No. APFSOS II/SP/2009/05. Nepal Forestry Outlook Study

24. Ohler, F. The Impact of Leasehold Forestry on Livelihoods in Nepal, Investment Centre, FAO, May 2012

25. Regmi, D. and Koirala, P.N., 8 June 2013. Western Upland Poverty Alleviation Project (WUPAP) – Component: Leasehold Forest and Non Timber Forest Programme, SWOT Analysis

26. Shapiro, B. 2001. Voices from the Field. Women's access to land and other natural resources in Nepal. Woman's Resource Access Programme

27. Sharma, Bishnu Prasad, June 2011. The Welfare Impacts of Leasehold Forestry in Nepal. Working Paper No. 61-11, South Asian Network for Development and Environmental Economics (SANDEE)

28. Singh, B.K. and Chapagain, D.P. Trends in forest ownership, forest resources tenure and institutional arrangements: are they contributing to better forest management and poverty reduction? Community and leasehold forestry for the poor: Nepal case study (undated)

29. Thapa, Y.B., Goutam, K.R. and Bhattrai. N., 2009. A Pro-Poor Perspective to Common Forest Management: Evidences from Leasehold Forestry in Central Nepal